AS

AQA

Biology

Steve Potter

Philip Allan Updates, an imprint of Hodder Education, part of Hachette Livre UK, Market Place, Deddington, Oxfordshire OX15 0SE

Orders
Bookpoint Ltd, 130 Milton Park, Abingdon, Oxfordshire OX14 4SB
tel: 01235 827720
fax: 01235 400454
e-mail: uk.orders@bookpoint.co.uk

Lines are open 9.00 a.m.–5.00 p.m., Monday to Saturday, with a 24-hour message answering service. You can also order through the Philip Allan Updates website: www.philipallan.co.uk

© Philip Allan Updates 2008

ISBN 978-0-340-95764-6

Impression number   5 4 3 2
Year   2012   2011   2010   2009   2008

Printed in Italy.

Hachette Livre UK's policy is to use papers that are natural, renewable and recyclable products and made from wood grown in sustainable forests. The logging and manufacturing processes are expected to conform to the environmental regulations of the country of origin.

P3187

Contents

Introduction

About this book

This textbook is written specifically for students following the AQA AS Biology course. The topics are in broadly the same order as in the specification, however some items have been reordered to give a more coherent structure.

Chapters 1–8 deal with the content of Module 1. The first four chapters cover the key topics of:
- biological molecules
- cell structure
- the nature and properties of enzymes
- transport across membranes

All of these topics are set in the context of food and the digestive system.

Chapters 5 and 6 deal with the heart, circulation, lungs and breathing.

Chapters 7 and 8 cover aspects of disease of the digestive, circulatory and breathing systems, together with the way in which the body responds to disease conditions.

Chapters 9–16 cover the content of Module 2. Chapters 9–13 deal with different sources of variation and how variation can be inherited. They also consider the consequences of variation, such as how:
- variation in the proteins produced by a cell affect the functioning of the cell
- variation in size affects surface area and volume and, as a further consequence, how gas exchange and transport take place in different organisms

Chapter 14 considers how genetic variation can lead to evolution, while Chapter 15 looks at how different organisms can be classified.

Finally, Chapter 16 highlights the biodiversity that exists on the planet and the impact that human activity is having on that biodiversity.

Each chapter begins with a chapter outline and a brief introduction. Margin comments accompany the text. Some provide extra information to help clarify a point without interrupting the flow of text (identified by the symbol ◀. Others are examiner hints on what to do and what not to do in unit tests. These are identified by the symbol ℮.

Feature boxes are included to provide extra detail or to give information about applications of a topic.

Some of these features relate to How Science Works and may show how:
- the work of one research team often depends on previous research findings
- scientists can communicate their research findings
- correlational evidence does not constitute proof
- the way in which careful design of an investigation allows cause and effect to be inferred, if not proved
- the way in which research is refined to show cause and effect

These boxes are identified with the following symbol:

The main content of each chapter is followed by a comprehensive summary; this would be a good place to start your revision of that topic.

Each chapter ends with questions designed to test your understanding of that topic. Multiple-choice questions are followed by longer, structured examination-style questions. Some of these test aspects of How Science Works.

The unit tests

Terms used in the unit tests

You will be asked precise questions in the examinations, so you can save a lot of valuable time as well as ensuring you score as many marks as possible by knowing what is expected.

Terms most commonly used are explained below.

Describe
This means exactly what it says — 'tell me about…' — and you should not need to explain why.

Explain
Here you must give biological reasons for why or how something is happening.

Complete
You must finish off a diagram, graph, flow chart or table.

Draw/plot
This means that you must construct some type of graph. For this, make sure that:
- you choose a scale that makes good use of the graph paper (if a scale is not given) and does not leave all the plots tucked away in one corner
- you plot an appropriate type of graph — if both variables are continuous variables, then a line graph is usually the most appropriate; if one is a discrete variable, then a bar chart is appropriate
- you plot carefully using a sharp pencil and draw lines accurately

From the...

This means that you must use only information in the diagram/graph/photograph or other forms of data.

Name

This asks you to give the name of a structure/molecule/organism etc.

Suggest

This means 'give a plausible biological explanation for' — it is often used when testing understanding of concepts in an unfamiliar situation.

Compare

In this case, you have to give similarities *and* differences.

Calculate

This means add, subtract, multiply, divide (do some kind of sum!) and show how you got your answer — always show your working!

'Do's and 'don't's

When you finally open the test paper, it can be quite a stressful moment. For example, you may not recognise the diagram or graph used in question 1. It can be quite demoralising to attempt a question at the start of an examination if you are not feeling very confident about it. So:

- *do not* begin to write as soon as you open the paper
- *do not* answer question 1 first, just because it is printed first (the examiner did not sequence the questions with your particular favourites in mind)
- *do* scan all the questions before you begin to answer any
- *do* identify those questions about which you feel most confident
- *do* answer first those questions about which you feel most confident, regardless of order in the paper
- *do* read the question carefully — if you are asked to explain, then explain, don't just describe
- *do* take notice of the mark allocation and don't supply the examiner with all your knowledge of osmosis if there is only 1 mark allocated (similarly, you will have to come up with four ideas if 4 marks are allocated)
- *do* try to stick to the point in your answer (it is easy to stray into related areas that will not score marks and will use up valuable time)
- *do* take care with:
  - drawings — you will not be asked to produce complex diagrams, but those you do produce must resemble the subject
  - labelling — label lines must touch the part you are required to identify; if they stop short or pass through the part, you will lose marks
  - graphs — draw small points if you are asked to plot a graph and join the plots with ruled lines or, if specifically asked for, a line or smooth curve of best fit through all the plots
- *do* try to answer all the questions

Chapter 1

We are what we eat

This chapter covers:
- the range of biological molecules that occur within living organisms
- the structure and function of some biological molecules:
  - carbohydrates (monosaccharides, disaccharides and polysaccharides)
  - proteins
  - lipids (triglycerides and phospholipids)
- methods of identifying biological molecules:
  - biochemical tests and ways of quantifying them

The foods that make up our diet supply the nutrients needed by our bodies. What, then, are nutrients and why do we need them?

The main classes of nutrient are carbohydrates, lipids, proteins, vitamins, mineral ions, water and dietary fibre. Foods are made from parts of other organisms and the nutrients are contained within the cells of those organisms. To obtain nutrients, food is first chewed to break open the cells. The contents are then digested and the products of digestion are absorbed.

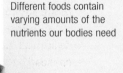

Different foods contain varying amounts of the nutrients our bodies need

Many nutrients are the biological molecules that we need to build our own cells. Some are respiratory substrates — they are respired, aerobically or anaerobically, to release energy that is used to synthesise ATP. The energy held in ATP molecules is used to 'drive' the many reactions that occur within a cell, including the synthesis of large molecules and cell components. Other nutrients are not used directly in synthesis or as respiratory substrates. For example, many vitamins act as coenzymes and help to catalyse reactions. Dietary fibre is a mixture of several complex carbohydrates, none of which is digested or absorbed. Yet dietary fibre is essential in our diet; among other functions, it helps to protect against colon cancer.

Many types of biological molecule are essential to build our bodies and to keep them functioning properly.

What are carbohydrates and why do we need them?

All carbohydrates contain the elements carbon, hydrogen and oxygen. The hydrogen and oxygen atoms in a carbohydrate molecule are present in the ratio of two hydrogen atoms to one oxygen atom (e.g. glucose, $C_6H_{12}O_6$ and maltose, $C_{12}H_{22}O_{11}$). Carbohydrates range from very small molecules containing only 12 atoms, to very large molecules containing thousands of atoms.

Carbohydrates have a range of functions:
- Glucose is the main respiratory substrate of most organisms.
- Storage carbohydrates include:
  - starch in plants
  - glycogen in animals
- Structural carbohydrates include:
  - cellulose, which is the main constituent of the primary cell wall of plants
  - chitin, which occurs in the cell walls of fungi and in the exoskeletons of insects
  - peptidoglycan, which occurs in bacterial cell walls

In plasma membranes, carbohydrates are found combined with proteins to form glycoproteins. Glycoproteins often have antigenic properties. They act as markers for the immune system, which can differentiate between 'self' antigens (those that are normally found in the body) and foreign or 'non-self' antigens. The presence of 'non-self' antigens stimulates an immune response, such as the production and secretion of antibodies to destroy the antigen. More detail on the immune response can be found in Chapter 8.

What different types of carbohydrate are there?

Monosaccharides are the simplest carbohydrates. A monosaccharide molecule can be thought of as a single unit. Other, more complex, carbohydrates have two or more such units joined together.

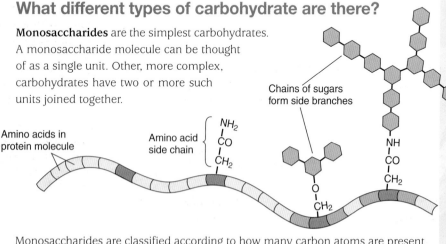

Figure 1.1 Carbohydrate chains combine with protein molecules in the plasma membrane to form glycoproteins

Chains of sugars form side branches

Amino acids in protein molecule

Amino acid side chain

NH₂
CO
CH₂

NH
CO
CH₂

O
CH₂

Monosaccharides are classified according to how many carbon atoms are present in the molecule:
- a **triose** monosaccharide has three carbon atoms
- a **tetrose** monosaccharide has four carbon atoms
- a **pentose** monosaccharide has five carbon atoms
- a **hexose** monosaccharide has six carbon atoms

◀ The biological molecules that make up our bodies are organic molecules. All organic molecules contain both carbon and hydrogen. For example, glucose ($C_6H_{12}O_6$, a monosaccharide carbohydrate) and glycine ($C_2H_5O_2N$, an amino acid) are both organic molecules. A molecule of carbon dioxide (CO_2) contains carbon but not hydrogen, and so is inorganic.

Sedoheptulose is a monosaccharide with seven carbon atoms in its molecule. It is an intermediate in the reactions of the Calvin cycle in the light-independent reactions of photosynthesis. D-glycero-d-manno-octulose has a molecule containing eight carbon atoms. It is found only in avocados and its ◀ function is uncertain.

Figure 1.2 shows the structures of four hexose sugars — α-glucose, β-glucose, galactose and fructose.

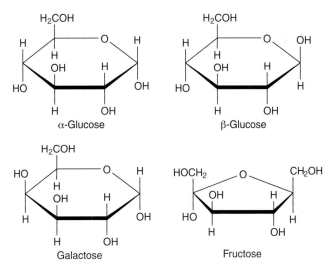

Figure 1.2 The ring forms of α-glucose, β-glucose, galactose and fructose — carbon atoms are found at the 'angles'

◀ It is easy to think that all the atoms of ring molecules lie in the same plane, but this is not so. The diagram below shows a three-dimensional representation of the atoms in α-glucose.

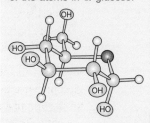

You do have to be able to recall the structure of α-glucose. However, you do not have to remember the position of every hydrogen and oxygen atom, only the simplified structure shown in Figure 1.3. This shows the overall shape of the molecule, the position of each carbon atom and the hydrogen and oxygen atoms attached to carbon atoms 1 and 4.

Figure 1.3 A simplified representation of the structure of α-glucose

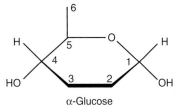

α-Glucose

◀ Glucose, galactose and fructose are known as reducing sugars. They reduce the copper ions (Cu^{2+}) in Benedict's solution to form copper(I) oxide (Cu_2O). This is insoluble and forms a red precipitate — the recognisable result of the test (see page 16).

Although you only need to be able to recall the structure of α-glucose, simplified structures of galactose and fructose are shown in Figure 1.4. Being aware of these will help you to understand the structures of some disaccharide sugars.

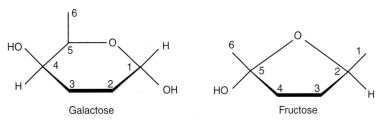

Galactose   Fructose

Figure 1.4 A simplified representation of the structures of galactose and fructose

Disaccharide carbohydrate molecules are made by two monosaccharide molecules joining together. For example, a molecule of:

- **maltose** is derived from two α-glucose molecules
- **sucrose** is derived from an α-glucose molecule and a fructose molecule
- **lactose** (milk sugar) is derived from a β-glucose molecule and a galactose molecule

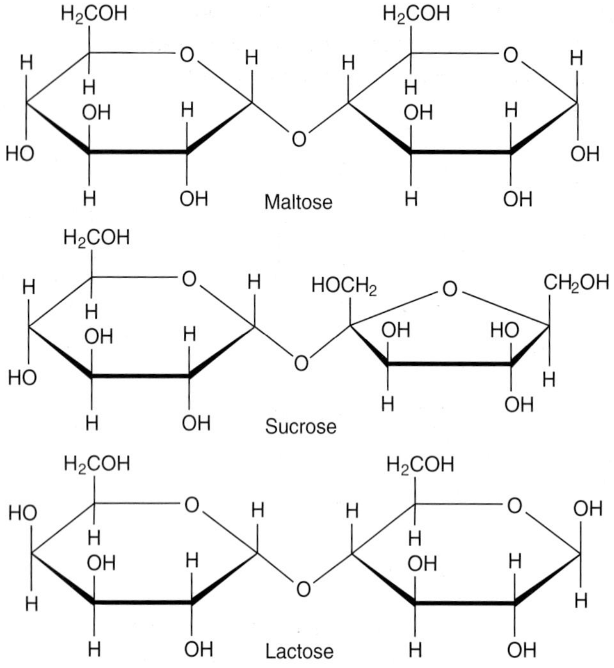

Figure 1.5 The structures of three disaccharides

In each of these examples, two hexose monosaccharides react to form a disaccharide molecule. As the formula of a hexose is $C_6H_{12}O_6$, you might expect the formula of the disaccharides to be $C_{12}H_{24}O_{12}$. In fact, the formula is $C_{12}H_{22}O_{11}$. A molecule of water (H_2O) is formed from a hydroxyl group from one monosaccharide and a hydrogen atom from the other (Figure 1.6). This allows a bond to be formed between the two monosaccharide units to make a disaccharide

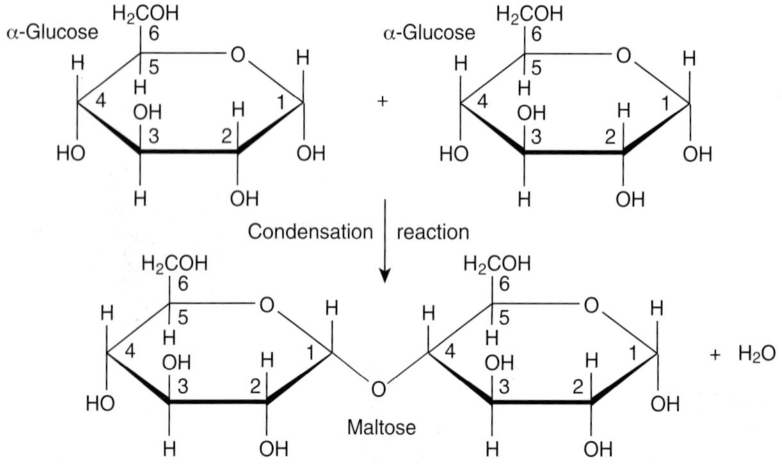

Figure 1.6 Two molecules of α-glucose are joined to form a molecule of maltose (a disaccharide)

The process shown in Figure 1.6 is called **condensation**. The bond formed is called a **glycosidic bond**. It is formed between carbon atom number 1 of one α-glucose molecule and carbon atom number 4 of another α-glucose molecule. The full name of the bond is, therefore, an α-1,4-glycosidic bond.

Condensation does not just occur in the formation of disaccharides, but also in the formation of polysaccharides and other large molecules.

The reverse process is **hydrolysis** of the disaccharide (or other molecule). This involves 'putting back' the water that was removed during condensation and splitting the molecule into its component, smaller molecules (Figure 1.7).

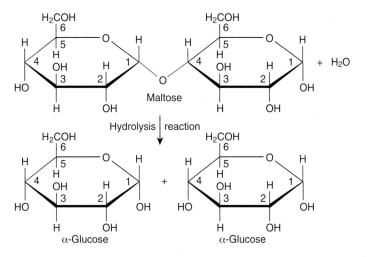

Figure 1.7 Hydrolysis of maltose

◄ Hydrolysis involves the addition of a molecule of water to break the bond formed during condensation.

Lactose intolerance is a condition that arises because many people cannot digest lactose (found in milk and other dairy products) into its constituent monosaccharides, glucose and galactose. Digestion of lactose is carried out by the enzyme **lactase**, which is present in all young humans. By the age of about 4 years, by which time breast-feeding has usually ended, many children lose the ability to make lactase. However, this is not true of all human populations. For example, most Europeans and North Americans retain the ability to make lactase into adulthood.

e The specification does not require you to know disaccharide structures in detail. However, you should know which monosaccharides make up which disaccharides and be able to explain the processes of condensation and hydrolysis.

Figure 1.8 Levels of lactose intolerance in different countries and among different ethnic groups within countries

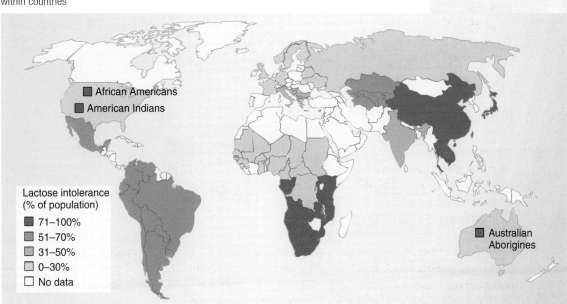

■ African Americans
■ American Indians

■ Australian Aborigines

Lactose intolerance (% of population)
■ 71–100%
■ 51–70%
■ 31–50%
□ 0–30%
□ No data

Without lactase, the lactose remains undigested and cannot be absorbed into the bloodstream. Bacteria living in the gut switch their metabolism and begin to ferment the lactose, producing large amounts of gas. The gas causes bloating, cramps, flatulence and diarrhoea. Lactose-intolerant people must consume only small amounts of dairy produce because this is a major source of lactose.

Lactose intolerance is usually an inherited condition, but some cases of 'late-onset lactose intolerance' may be the result of intestinal infections. This type of lactose intolerance is often temporary, but may be permanent.

Polysaccharides are complex carbohydrates. Their molecules are made by many hundreds of monosaccharide molecules joining together by condensation links.

Box 1.1 Macromolecules, polymers and monomers

A macromolecule is just what the name suggests — it is a big molecule. Examples include proteins, starch, cellulose, glycogen and DNA. A polymer is also a large molecule — but it is not *just* large. A polymer molecule is made from many smaller, often identical, molecules called monomers. Besides being macromolecules, starch, glycogen, cellulose and proteins are polymers. Starch and glycogen are polymers of α-glucose, cellulose is a polymer of β-glucose and proteins are polymers of amino acids.

Figure 1.9 Linkages in amylose and the amylose helix

Starch is not a single compound but a mixture of **amylose** and **amylopectin**. Both are polymers of α-glucose, but the arrangement of the α-glucose monomers in these compounds is different. Amylose is a linear molecule containing many hundreds of α-glucose molecules joined by α-1,4-glycosidic bonds (as in maltose). As it is being formed, this long chain winds itself into a helix.

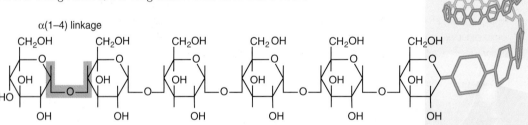

Box 1.2 Starch stains blue or black or blue-black with iodine

The helical structure of amylose allows a reaction between starch and iodine solution to occur. Rows of iodine atoms sit inside the amylose helix and interact with it, changing the light-absorbing properties of both, so that the complex appears blue. Starches in different plants have different proportions of amylose and amylopectin. This results in different shades of blue-black with the iodine test, because only the amylose reacts with iodine.

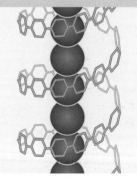

Amylopectin also has a linear 'backbone' of α-glucose molecules joined by α-1, 4-glycosidic bonds. In addition, there are side branches. At certain points along the chain, a glucose molecule forms an α-1,6-glycosidic bond with another glucose molecule.

◀ An enzyme that digests proteins is a *protease*; one that digests lipids is a *lipase*. An enzyme that digests starch is an *amylase* because starch is made from *amyl*ose and *amyl*opectin.

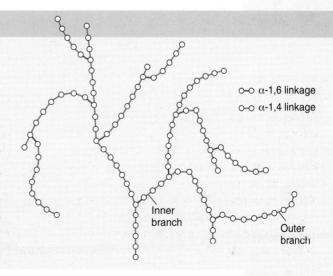

Figure 1.10 Structure of amylopectin

Starch is a plant storage carbohydrate. Since both amylose and amylopectin are compact molecules, many α-glucose molecules can be stored in a small space, without affecting cell metabolism. If glucose were stored without being converted to starch, it would create a negative water potential within the cytoplasm. This would draw water, by osmosis, from neighbouring cells and from organelles within the cell. Starch is insoluble and produces none of these effects. In addition, because starch is insoluble, the molecules cannot move out of cells — they remain in storage organs. The branched nature of amylopectin means that there are many 'ends' to the molecule. Therefore, starch can be quickly hydrolysed (by enzymes acting at the ends of the chains) to release glucose for respiration.

Box 1.3 Glycogen

Glycogen is a storage carbohydrate in animal cells. The molecular structure is similar to that of the amylopectin component of starch but there are more α-1,6 links, so the molecule is more highly branched. This means that glycogen can be hydrolysed more quickly to release glucose than starch. This is important because animals have a higher metabolic rate than plants and need to release energy more quickly to 'drive' their metabolic processes.

o—o α-1,6 linkage
o—o α-1,4 linkage

Inner branch

Outer branch

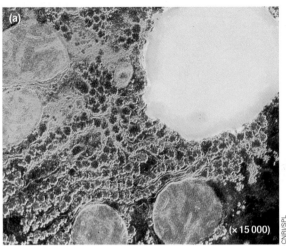
(×15 000)

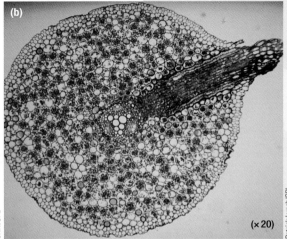
(×20)

Who needs proteins?

Proteins are important substances that are needed to form all living cells. Their molecules contain the elements carbon, hydrogen and oxygen (as do carbohydrates), but they also contain nitrogen and most contain sulphur. Protein molecules are polymers of amino acids and are, therefore, also macromolecules. Even so, they vary enormously in size. The smallest protein molecules contain fewer than 100 amino acids; the largest contain several thousand. Proteins have a range of functions — for example, they are important in:

- the structure of **plasma membranes** — protein molecules form ion channels, transport proteins and surface receptors for hormones, neurotransmitters and other molecules
- the **immune system** — antigen and antibody molecules are proteins (Chapter 8)
- the **enzymic control** of metabolism — all enzymes are proteins (Chapter 3)
- the structure of **chromosomes** — DNA is wound around molecules of the protein histone to form a chromosome (Chapter 10)

Although proteins are referred to as polymers, this is not strictly true. In a true polymer, all the monomers (molecules making up the polymer) are identical. This is true of amylose and amylopectin in starch (all the monomers are α-glucose) and of cellulose (all the monomers are β-glucose). In proteins, although all the monomers are amino acids, there are different amino acids in any given protein molecule. However, since all amino acids have the same basic structure, it is acceptable to refer to a protein molecule as a polymer.

All amino acid molecules are built around a carbon atom to which is attached:

- a hydrogen atom
- an amino group ($-NH_2$)
- a carboxyl group ($-COOH$)
- an 'R' group, which represents the other atoms in the molecule, such as a single hydrogen atom, a hydrocarbon chain or a more complex structure

(a) False-colour transmission electron micrograph of a section through a liver cell containing stored glycogen. The glycogen appears red.
(b) Light micrograph of starch grains (stained purple) in buttercup root cells

Figure 1.11 The general structure of an amino acid

e You may be asked to draw the general structure of an amino acid.

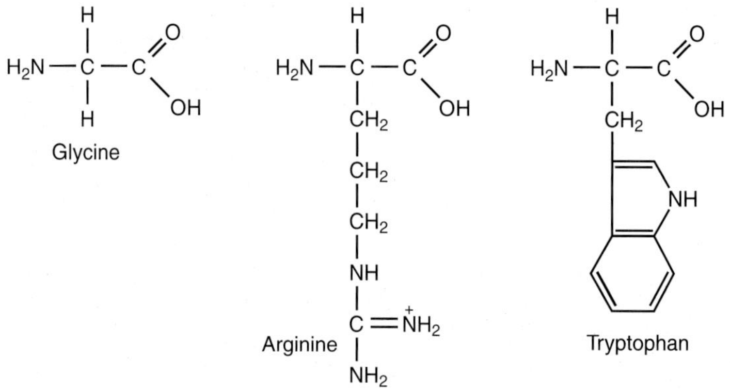

Figure 1.12 Structures of three amino acids

Glycine

Arginine

Tryptophan

Two amino acids can be joined together by condensation to form a **dipeptide**. A dipeptide can be enlarged into a **polypeptide** by condensation with more amino acid molecules. The bonds formed by condensation can be broken by hydrolysis.

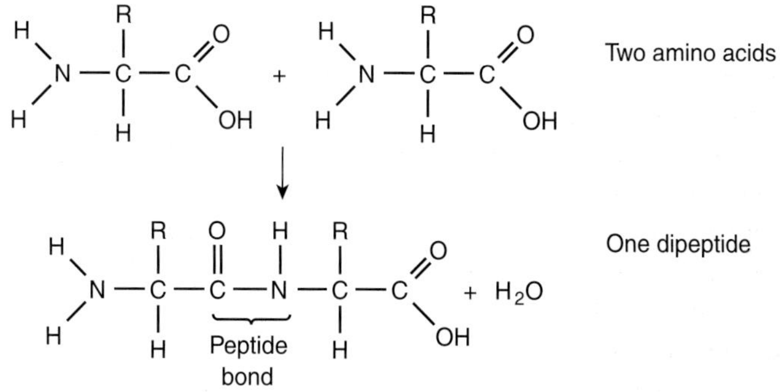

Figure 1.13 How a dipeptide is formed

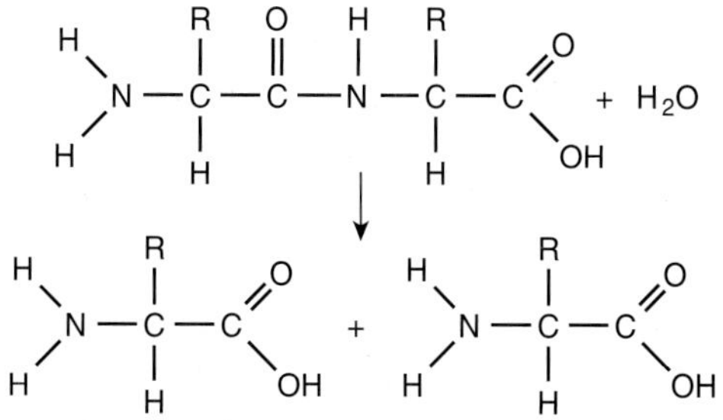

Two amino acids

One dipeptide

+ H_2O

Peptide bond

Figure 1.14 Hydrolysis of a dipeptide

+ H_2O

How is a protein molecule put together?

Many amino acids joined by peptide bonds form a polypeptide chain; this sequence of amino acids is the **primary structure** of the protein. Once formed, the polypeptide chain folds itself into a **secondary structure**, which is either an α-helix or a β-pleated sheet. The structures are held in place by hydrogen bonds that form between peptide bonds in adjacent parts of the amino acid chain. Both types of secondary structure can exist in different regions of the same polypeptide chain.

α-helix β-pleated sheet

Figure 1.15 The secondary structure of proteins — sections of the amino acid chain fold into either an α-helix or a β-pleated sheet

A protein molecule can have a **tertiary structure**. This involves further folding of the secondary structure and the formation of new bonds to hold the tertiary structure in place. These new bonds include:

- more hydrogen bonds — between the R-groups of some amino acids
- disulphide bridges — between amino acids with R-groups that contain sulphur
- ionic bonds — between amino acids with positively charged R-groups and amino acids with negatively charged R-groups

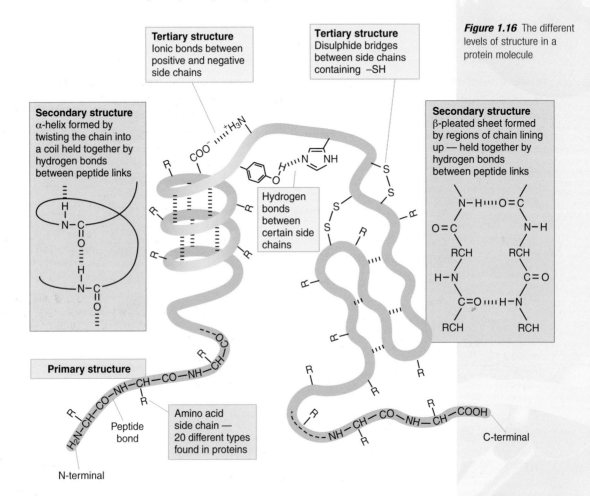

Tertiary structure
Ionic bonds between positive and negative side chains

Tertiary structure
Disulphide bridges between side chains containing –SH

Figure 1.16 The different levels of structure in a protein molecule

Secondary structure
α-helix formed by twisting the chain into a coil held together by hydrogen bonds between peptide links

Secondary structure
β-pleated sheet formed by regions of chain lining up — held together by hydrogen bonds between peptide links

Hydrogen bonds between certain side chains

Primary structure

Peptide bond

Amino acid side chain — 20 different types found in proteins

N-terminal

C-terminal

Each protein has a unique tertiary structure and, therefore, a unique configuration (shape):

- The primary structure of each protein is controlled genetically. This determines the type and position of each amino acid in the polypeptide chain.
- The secondary structure of the molecule is the consequence of its primary structure. Some sections of the primary structure form α-helices; others form β-pleated sheets.
- The nature of the secondary structure determines where ionic and hydrogen bonds and disulphide bridges form, i.e. it determines the tertiary structure and shape of the protein molecule.

Figure 1.17 The tertiary structure of a protein molecule

Because the tertiary structure of each protein is unique, each has a specific function. Some functions of proteins are described in Table 1.1.

Protein molecule	Effect of specific shape	Consequence
Enzyme	Active site binds only with certain molecules	Enzyme is specific — it will only catalyse a particular reaction
Insulin receptor in plasma membrane	Only insulin binds with this receptor	Insulin targets only cells with the receptor
Receptor in 'sweet' taste bud	Only binds with molecules with a shape that fits	These molecules taste sweet

Proteins are classified into two main groups, according to their molecular shapes. A **fibrous protein** has a tertiary structure that resembles a long string or fibre. Fibrous proteins are usually structural. Two examples are collagen (found in bone, cartilage and many other structures) and keratin (found in skin and nails). A **globular protein** has a tertiary structure that resembles a globule or ball. The proteins involved in controlling cellular metabolism — for example, enzymes and receptor proteins — are globular.

A few proteins have a **quaternary structure**. In these cases, two or more polypeptide chains are bonded together to form the final protein molecule. Haemoglobin consists of four polypeptide chains, two α-chains and two β-chains. A molecule of collagen consists of three polypeptide chains wound around each other.

Table 1.1 How a unique tertiary structure determines the properties of proteins

e You may be asked to explain how the primary structure of a protein determines its final function.

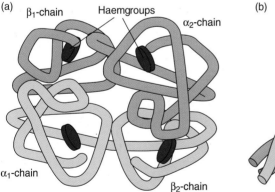
(a) β₁-chain   Haemgroups   α₂-chain   α₁-chain   β₂-chain

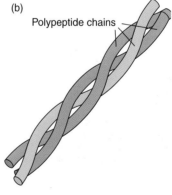
(b) Polypeptide chains

Figure 1.18 The quaternary structures of (a) haemoglobin and (b) collagen

Box 1.4 The insulin story

Today we take it for granted that people suffering from diabetes will be able to obtain human insulin, usually produced by genetically engineered bacteria. However, it took a number of scientists a long time to make this possible.

- In 1869, Paul Langerhans discovered clumps of tissue in the pancreas that are different from normal pancreatic tissue. These were later named the islets of Langerhans. He suggested that they may have some regulatory role in digestion.
- In 1889, Oskar Minkowski, knowing of Langerhans's work, removed the pancreas from a healthy dog. He noticed that, after the operation, flies swarmed around the dog's urine. On testing, the urine was found to contain sugar. This established a link between the pancreas and diabetes. At the time, diabetes was almost always fatal.

- In 1901, following on from Minkowski's work, Eugene Opie showed that diabetes occurred when the islets of Langerhans were destroyed, refining the link with diabetes to the islets of Langerhans, rather than the whole pancreas.
- Between 1901 and 1920, a number of scientists tried, unsuccessfully, to produce extracts of the islets of Langerhans.
- In 1921–22, after reading about the work of the other scientists, Frederick Banting and Charles Best (together with J. J. R. Macleod and James Collip) developed a method for obtaining an extract from the islets of Langerhans. They called the extract 'isletin' and showed that it could relieve the symptoms of diabetes. In a dramatic trial, they visited a hospital ward where there were 50 children in comas, dying from diabetes. They injected the children with the 'isletin' extract. Before they had reached the end of the ward, some of the first children they had injected were awakening from their comas.
- In 1922, scientists working for the drug company Eli Lilly made a major breakthrough in preparing a pure extract, and insulin (as it was renamed) went on sale soon afterwards.
- In 1958, Fred Sanger discovered the amino acid sequence in the insulin molecule. It was the first protein molecule to have its amino acid sequence determined.
- In 1969, Dorothy Crowfoot-Hodgkin determined the tertiary structure of the insulin molecule.

The amino acid sequence of insulin

Many of these discoveries depended on the scientists knowing of earlier work. By publishing their findings in journals, scientists make it possible for those who come to work on the problem later to have a new 'starting point' for their research. The insulin story shows how the understanding of the role of the pancreas in diabetes became increasingly refined as more evidence became available.

However, there are ethical issues involved. Much of the early research involved experimentation on dogs, calves and oxen. Is this acceptable? The production of pure insulin by Eli Lilly made huge profits for the company, but it also saved many lives. Is it acceptable for drug companies to make such profits from human suffering?

Finally, the insulin story shows how scientific advances improved the quality of life of millions of diabetes sufferers. Without insulin, they would almost certainly have died prematurely.

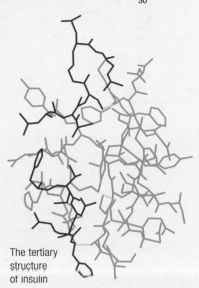

The tertiary structure of insulin

Why eat lipids?

Lipids are a varied group of compounds that include triglycerides, phospholipids and waxes. Unlike proteins and polysaccharides, lipids are *not* polymers. Their molecules are relatively small. The feature they share is that they are all esters of fatty acids and alcohols. **Waxes** are formed from fatty acids and long-chain alcohols. This structure makes them so insoluble in water that they can act as water repellents — for example, in coating birds' feathers. **Phospholipids** are one of the basic components of cell membranes. **Triglycerides** have several functions including:

- **respiratory substrate** — a molecule of triglyceride yields over twice as many molecules of ATP as a molecule of glucose
- **thermal insulation** — the cells of adipose tissue found under the skin contain large amounts of triglycerides and give good thermal insulation
- **buoyancy** — lipids are less dense than water (oil floats on water) and large amounts of lipid reduce the density of an animal
- **waterproofing** — the oils secreted by some animals onto their skins are tri-glycerides

Nearly all lipids contain only carbon, hydrogen and oxygen. A lipid molecule contains much less oxygen than a carbohydrate molecule of similar molecular mass. Therefore, more oxygen is needed to respire the lipid molecule.

> Some of the lipids found in the myelin sheath that surrounds nerve cells are sphingolipids. These lipids are unusual in containing nitrogen as well as carbon, hydrogen and oxygen.

More about triglycerides

A triglyceride molecule is an ester formed from one molecule of **glycerol** (an alcohol containing three carbon atoms) and three **fatty acid** molecules.

A fatty acid molecule consists of a covalently bonded **hydrocarbon chain**, at the end of which is a carboxyl group, which has acidic properties. The hydrocarbon chain is non-polar. This means that it has no charge. The carboxyl group is ionic and dissociates in solution to form COO^- and H^+ (hydrogen ion). The hydrogen ions released make the solution acidic.

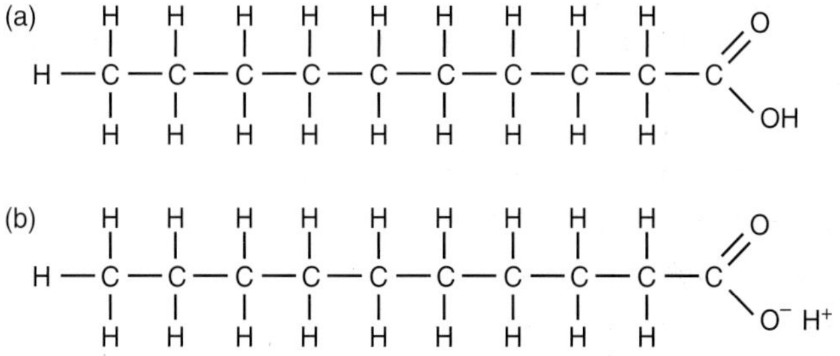

> Glycerol is a polyhydroxy alcohol — it contains more than one hydroxyl (–OH) group. Ethanol, the alcohol in beer and wine, has the formula C_2H_5OH. It is a monohydroxy alcohol — it contains only one hydroxyl group:

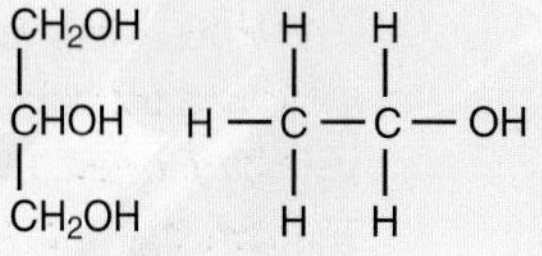

Figure 1.19
(a) The structure of a typical fatty acid molecule
(b) The acidic nature of fatty acids

The nature of the hydrocarbon chains in fatty acids can differ in two main ways:
- The number of carbon atoms in the chains can vary.
- Hydrocarbon chains with the same number of carbon atoms can have different numbers of hydrogen atoms. This is because of the nature of the bonding between the carbon atoms in the chain. If all the carbon–carbon bonds in the hydrocarbon chain are single bonds, the fatty acid is a **saturated fatty acid**. If one

of the carbon–carbon bonds is a double bond, then it is a **monounsaturated fatty acid**. If more than one carbon–carbon bond is a double bond, then the fatty acid is a **polyunsaturated fatty acid**. For example, the hydrocarbon chain in stearic acid contains no double carbon–carbon bonds. It is a saturated fatty acid (Figure 1.20 (a)). Oleic acid has the same number of carbon atoms as stearic acid, but is monounsaturated (Figure 1.20(b)); linoleic acid also has the same number of carbon atoms as stearic acid but is polyunsaturated (Figure 1.20(c)).

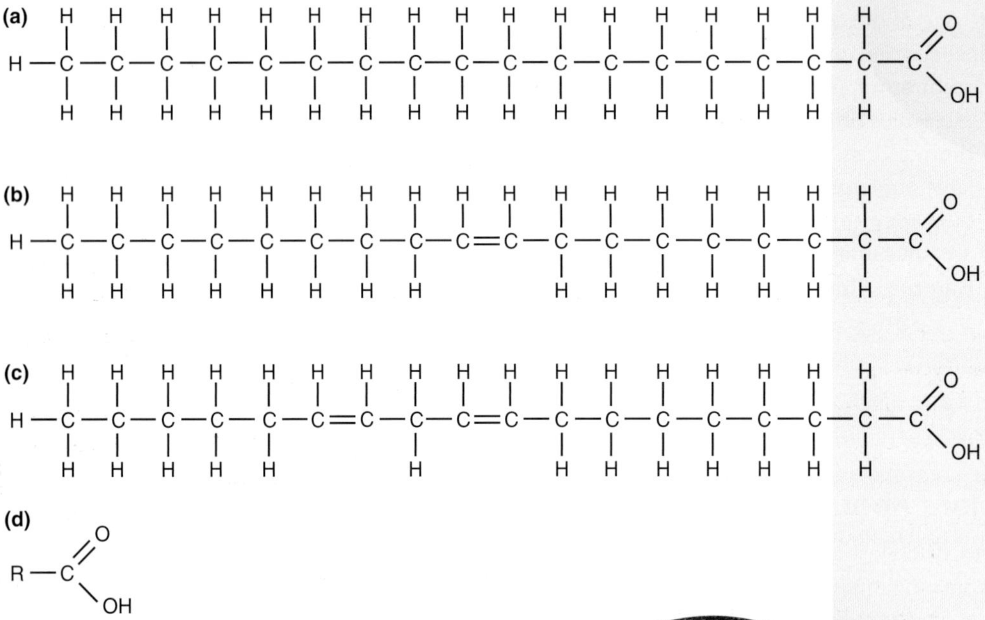

Figure 1.20
(a) Stearic acid
(b) Monounsaturated oleic acid
(c) Polyunsaturated linoleic acid
(d) The generalised structure of a fatty acid; 'R' represents the hydrocarbon chain

Box 1.5 High in polyunsaturates

You may have seen cooking oils, margarines and other spreads advertised as 'high in polyunsaturates'. This means that the lipids in the product contain a high proportion of polyunsaturated fatty acids. Polyunsaturated fatty acids help to prevent cholesterol being laid down in the linings of arteries (atherosclerosis) and so help to prevent heart disease.

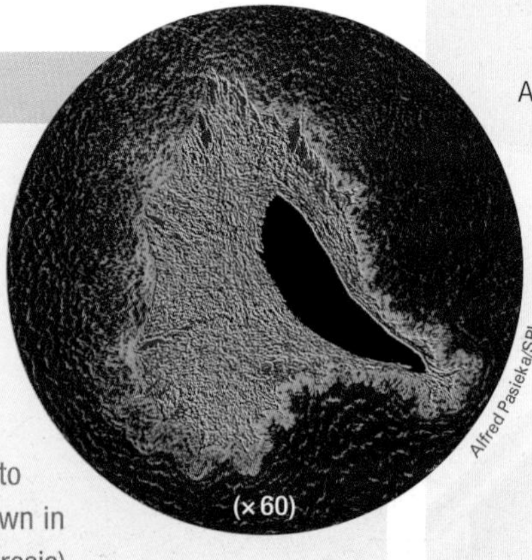

Atherosclerosis in an artery; eating polyunsaturated fatty acids can help to prevent this

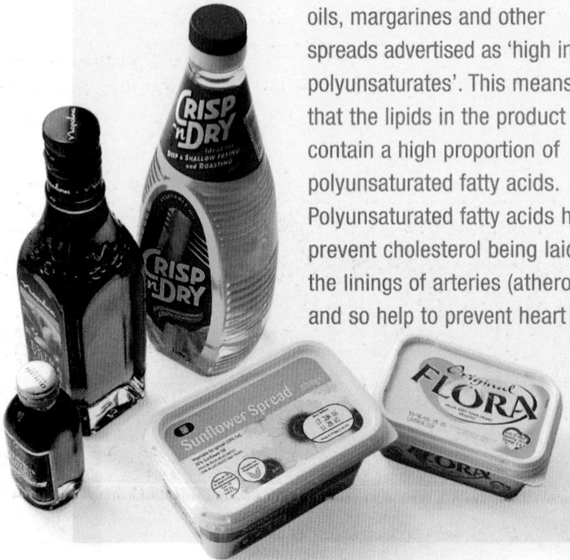

Alfred Pasieka/SPL

(×60)

These products contain a high proportion of polyunsaturated or monounsaturated fatty acids

You may be asked to distinguish between saturated and unsaturated fatty acids or to complete diagrams showing saturated or unsaturated fatty acids.

The tests are summarised in Table 1.2.

Step in test	Reducing sugar	Non-reducing sugar
Heat with Benedict's solution	Red precipitate; reducing sugar present — no further steps needed	No change; solution remains blue; reducing sugar absent — proceed
Boil with hydrochloric acid for 5 minutes		Acid hydrolyses non-reducing sugar molecules
Neutralise with sodium carbonate solution		Benedict's solution reacts only in neutral or alkaline conditions
Re-test with Benedict's solution		Red precipitate
Conclusion	Reducing sugar present originally	Non-reducing sugar present originally; reducing sugars now present were formed by acid hydrolysis of a non-reducing sugar

Table 1.2 Testing for reducing and non-reducing sugars

◀ When carrying out the test, the test sample should be heated with Benedict's solution in a water bath at 85°C or higher.

It is important to note that the Benedict's test does not distinguish between different reducing sugars. It is *not* a test for glucose — or galactose or any individual sugar. To distinguish between sugars, enzyme-based tests are used.

Testing for lipids

The test for a lipid is based on the fact that lipids are soluble in organic solvents such as ethanol, but insoluble in water. This test is called the **emulsion test** and is carried out as follows.

- Shake the test sample with ethanol, in a clean, dry test tube.
- Filter the mixture (if necessary).
- Pour the filtrate into water.

Any lipid in the filtrate will not dissolve in the water. It will form an emulsion that makes the liquid appear milky white.

Testing for proteins

There are several biochemical tests for proteins, but the simplest, and often the most reliable, is the **Biuret test**. In this test, a protein in an alkaline solution reacts with copper ions to produce a mauve/purple colour. There are two ways of carrying out the test:

- Method 1:
  - The test solution is mixed with sodium hydroxide solution in a test tube.
  - A few drops of 1% copper(II) sulphate solution are added.
  - The mixture is allowed to stand for a few minutes to allow the colour to develop fully.
- Method 2:
  - The test solution is mixed with Biuret solution (which contains copper ions in an alkaline solution).
  - The mixture is allowed to stand for a few minutes to allow the colour to develop fully.

Andrew Lambert Photography/SPL

Lipids produce a milky white emulsion when dispersed in water

The Biuret test can be made semi-quantitative in the following way:

- Add 2 cm³ Biuret solution to 5 cm³ 0.1 mol dm⁻³ protein solution in a test tube.
- Leave for 10 minutes.
- Transfer a sample to a cuvette.
- Insert the cuvette into a colorimeter and measure the percentage transmission of light.
- Repeat with protein solutions of different concentrations (0.2, 0.4, 0.6, 0.8, 1.0 mol dm⁻³) and with distilled water.
- Plot a graph of percentage transmission against concentration. This is called a **calibration curve**.
- Repeat with the test solution.
- Estimate the concentration of protein in the test solution from the calibration curve.

An alkaline protein solution produces a purple ring after addition of copper ions

Martyn F. Chillmaid/SPL

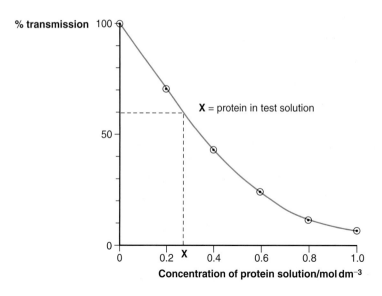

Figure 1.25 Estimating the percentage of protein in a test solution from the calibration curve

Box 1.6 How a colorimeter works

The principle behind a colorimeter is that when a sample of liquid is illuminated, not all the light is transmitted (passes through) — some is absorbed. A small container (cuvette) containing the test liquid is placed in the colorimeter and illuminated. The light that passes through the liquid is detected by a photocell. The instrument converts this into one of two readings:

- percentage transmission — the percentage of the light that passes through the sample
- absorbance — a measure (but *not* a percentage) of how much light is absorbed by the sample

There are other techniques that can be used to analyse foods. Biosensors can be used to provide assays of specific sugars (rather than just reducing sugars and non-reducing sugars) as well as of proteins and amino acids within proteins. Chromatography is still used to separate mixtures of amino acids. These are just two examples; there are many more.

Summary

Carbohydrates

- Carbohydrate molecules contain the elements carbon, hydrogen and oxygen only. The ratio of hydrogen atoms to oxygen atoms is 2:1.
- Monosaccharides are carbohydrates. The atoms are arranged in a single ring-like structure.
- The formula of α-glucose and all other hexose monosaccharides is $C_6H_{12}O_6$.
- The structure of α-glucose is:

α-Glucose

- Two monosaccharides can be joined by condensation to form a disaccharide. In the formation of maltose, two α-glucose molecules are joined with the loss of a molecule of water (H_2O). The formula of maltose is $C_{12}H_{22}O_{11}$.
- The bond joining the two molecules of a-glucose is an α-1,4-glycosidic bond.
- Lactose is made when α-glucose and galactose are joined in a condensation reaction.
- Sucrose is made when α-glucose and fructose are joined in a condensation reaction.
- Polysaccharides are formed when many monosaccharide molecules join by condensation. Starch contains two polymers of α-glucose — amylose and amylopectin..
- Starch and glycogen are storage carbohydrates. They have compact molecules that enable much glucose to be stored in a small place. They are insoluble, which means that they have no osmotic effects within the cell and do not move from the cell.

Proteins

- Amino acids contain carbon, hydrogen, oxygen and nitrogen. They have the following general structure:

- Amino acids can be joined by condensation. The bond between two amino acids is a peptide bond. A large number of amino acids joined in this way form a polypeptide.
- Proteins are polymers of amino acids. They have several levels of structure:
  - the primary structure is the sequence of amino acids in a polypeptide chain
  - the secondary structure is determined by the folding of the primary structure into either an α-helix or a β-pleated sheet; these structures are held in shape by hydrogen bonds
  - the tertiary structure is determined by the further folding of the secondary structure into either a fibrous or globular shape; these structures are held in place by further hydrogen bonds, disulphide bridges and ionic bonds
  - some have a quaternary structure in which two or more polypeptide chains, each with a tertiary structure, are bonded together; a haemoglobin molecule consists of four polypeptide chains bonded together

Lipids

- A triglyceride molecule is an ester of three fatty acid molecules and one glycerol molecule; the ester bonds are formed by condensation.

- Fatty acid molecules can be either saturated (all carbon–carbon bonds are single), monounsaturated (one carbon–carbon double bond) or polyunsaturated (more than one carbon–carbon double bond).
- A phospholipid molecule consists of two fatty acids and a phosphate group bonded to a molecule of glycerol. The phosphate group gives the molecule a hydrophilic 'head' and the fatty acids give the molecule hydrophobic 'tails'.
- Phospholipid bilayers are the basis of biological membranes.

Identifying biological molecules

- Reducing sugars react with Benedict's solution when heated to give a yellow/orange/red precipitate.
- Non-reducing sugars must first be hydrolysed by boiling with HCl and then neutralised before they will react with Benedict's solution.
- Proteins react with Biuret reagent to give a mauve/purple colour.
- The emulsion test for lipids produces a milky-white colour in water.

Questions

Multiple-choice

1 Hexoses are:

  A disaccharides with molecules that contain six carbon atoms

  B monosaccharides with molecules that contain six oxygen atoms

  C monosaccharides with molecules that contain six carbon atoms

  D disaccharides with molecules that contain six oxygen atoms

2 The main advantage of the high level of branching in a molecule of amylopectin is that:

  A the many 'ends' allow rapid hydrolysis

  B much can be stored in a small space

  C there are no osmotic effects

  D it is insoluble

3 Lactose intolerance is due to:

  A a lack of lactose in the diet

  B the inability to produce the enzyme lactase

  C over-production of the enzyme lactase

  D too much lactose in the diet

4 The secondary structure of a protein can be:

  A a globular or a fibrous structure

  B a specific sequence of amino acids

  C a dipeptide

  D an α-helix or a β-pleated sheet

5 Condensation involves:

  A the creation of new bonds with the addition of a molecule of water

  B the creation of new bonds with the loss of a molecule of water

  C the breaking of existing bonds with the addition of a molecule of water

  D the breaking of existing bonds with the loss of a molecule of water

6 A food gives a positive result when tested with the Biuret test and an initial negative result when tested with Benedict's solution. Following acid hydrolysis and neutralisation, the food gave a negative test with Benedict's solution. The food contains:

  A protein and a non-reducing sugar

  B protein and a reducing sugar

  C protein, a reducing sugar and a non-reducing sugar

  D protein only

7 In a saturated fatty acid:

  A there are only single bonds between carbon atoms

  B there is one double bond between carbon atoms

  C there is one triple bond between carbon atoms

  D there is more than one double bond between carbon atoms

8 Phospholipids form bilayers in water because:

  A the hydrophilic head is repelled by the water and the hydrophobic tail is attracted by it

  B the hydrophilic head is attracted by the water and the hydrophobic tail is repelled by it

C both the hydrophilic head and the hydrophobic tail are attracted by the water

D both the hydrophilic head and the hydrophobic tail are repelled by the water

9 When heated with Benedict's solution, sucrose does not cause a colour change because it is:

A a reducing sugar

B a disaccharide

C a non-reducing sugar

D a compound sugar

10 A triglyceride molecule is an ester of:

A three fatty acids and ethanol

B two fatty acids and glycerol

C two fatty acids and ethanol

D three fatty acids and glycerol

Examination-style

1 The diagram shows two amino acids:

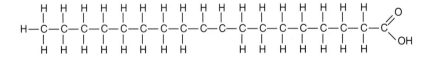

(a) (i) Copy the diagram and indicate the amino group on each amino acid. *(1 mark)*

(ii) Draw another diagram to show how these two amino acids could form a dipeptide. *(2 marks)*

(iii) Name both the process involved in the formation of the dipeptide and the type of bond formed. *(2 marks)*

(b) What is the secondary structure of a protein? *(2 marks)*

Total: 7 marks

2 The diagram shows the arrangement of the atoms in a monounsaturated fatty acid:

$$H-C-C-C-C-C-C-C-C-C-C-C-C-C-C-C-C-C-C\diagdown_{OH}^{O}$$

(a) Copy the diagram and add any double carbon–carbon bonds that may exist in this molecule. *(2 marks)*

(b) A molecule of this fatty acid contains more carbon atoms than a molecule of α-glucose.

(i) Give *three* other differences between the two molecules. *(3 marks)*

(ii) Give *one* similarity between the two molecules. *(1 mark)*

Total: 6 marks

3 Five solutions labelled P, Q, R, S and T are known to be:
- amylase (an enzyme that digests starch to maltose)
- albumen (a protein)
- starch
- sucrose
- glucose

The following tests are carried out:
- Each solution is tested with iodine solution. This allows identification of solution S.
- The remaining four solutions are tested with Benedict's solution. This allows identification of solution P.
- The remaining three solutions are tested with Biuret reagent. Solutions R and T both turn purple. Solution Q can now be identified.

(a) Identify, giving reasons, solutions P, Q and S. (3 marks)
(b) How could solutions R and T be distinguished? (4 marks)

Total: 7 marks

4 A molecule of lactose is formed by the condensation of a molecule of β-glucose and a molecule of galactose. Both β-glucose and galactose have the formula $C_6H_{12}O_6$.
(a) Explain how β-glucose and galactose can both have the formula $C_6H_{12}O_6$ and yet be different substances. (2 marks)
(b) How many oxygen atoms are there in a molecule of lactose? Explain why this is the case. (3 marks)
(c) Galactose is a reducing sugar. What does this mean? (2 marks)

Total: 7 marks

5 (a) Describe how you would test a solution to see if it contained a protein. (2 marks)
(b) The diagram shows the structure of a protein molecule.

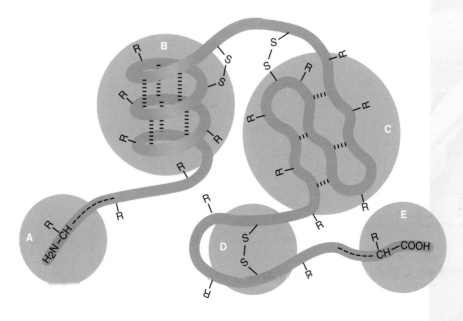

(i) Identify the region that represents:
- an α-helix (*1 mark*)
- a β-pleated sheet (*1 mark*)

(ii) Explain why the final shape of a protein molecule is determined by its primary structure. (*4 marks*)

Total: 8 marks

6 Lactose intolerance can cause sufferers quite serious abdominal pain. It is rare in children under the age of four years, but, in certain populations, it is common in adults. Scientists believe that lactose intolerance in adults was the normal condition and that lactose tolerance arose as a result of a mutation.

(a)(i) Give two other symptoms of lactose intolerance. (*2 marks*)

(ii) Explain the cause of the abdominal pain. (*2 marks*)

(b) Explain why some people become lactose intolerant as they grow older. (*2 marks*)

(c) Suggest why scientists believe that being lactose intolerant was the normal condition for the majority of people. (*3 marks*)

Total: 9 marks

Chapter 2

Molecules make cells

This chapter covers:

- the concepts of scale, magnification and resolution
- the way optical microscopes, transmission electron microscopes and scanning electron microscopes work
- how to interpret micrographs
- the limitations of light and electron microscopes
- the ultrastructure of an epithelial cell from the small intestine
- the ultrastructure of a bacterial cell
- cell fractionation as a technique for obtaining pure samples of different organelles for further study

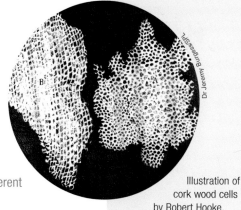

Illustration of cork wood cells by Robert Hooke, from observations of cork wood under a microscope

Many of the molecules we obtain from food are used to build the components of cells. Our present-day understanding of the ultrastructure of cells owes much to the work of pioneers in the field, for example Robert Hooke. His work was dependent on the work of Hans and Zacharias Janssen, who produced the first compound microscope.

A compound microscope is a microscope with two lenses.

In 1665, Robert Hooke used one of the first compound microscopes to examine cork. He found that it consisted of small units, which he called **cells**. Following this discovery, other researchers began to use microscopes to examine organisms — mainly plants. By 1839, two German botanists, Schleiden and Schwann, were so convinced by their findings that they put forward the **cell theory**, which stated that all organisms are composed of cells.

A modern version of the cell theory states that:

- a cell is the smallest independent unit of life — anything smaller than a cell is incapable of independent existence
- all organisms are cellular — some consist of just one cell; others contain billions
- cells arise from other cells by cell division — they cannot arise by spontaneous generation (be created from non-living materials)

Scientists used to believe that organisms could arise by spontaneous generation, i.e. they could be formed from non-living materials. As late as the 1600s, an eminent chemist put forward a 'recipe' for the spontaneous generation of mice!

Once again, we see that the work of a scientist is based on work carried out previously by other scientists.

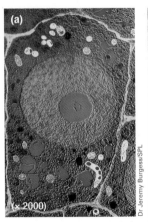
(×2000)

(×200)

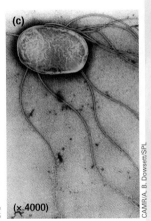
(×4000)

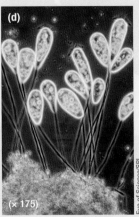
(×175)

Micrographs of (a) a cell in the root tip of a maize plant (b) columnar epithelial cells in the small intestine (c) a *Salmonella* bacterium (d) *Vorticella*

How big are organisms?

The largest single organism in the world at the moment is the giant redwood tree, which can grow up to 100 m in height. The largest animal is the blue whale, which can be up to 30 m long. Adult humans can be 2 m tall. Domestic cats are about 0.3 m long and goldfish 0.1 m. A housefly is about 0.01 m long. One of the largest human cells is an oocyte (egg cell), which has a diameter of 0.0002 m. Most animal cells are smaller than this, with a diameter of 0.00005 m. Typical bacterial cells have a length of 0.000005 m.

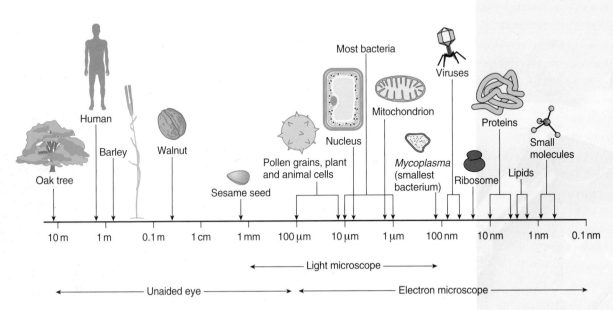

Figure 2.1 The range of sizes of living organisms and their components

All these zeros can be confusing. To make the numbers more manageable, different, but related, units are used. For example, it is easier to think of a housefly having a length of 1 cm rather than 0.01 m. Similarly, we can visualise an oocyte having a diameter of 0.2 mm more easily than 0.0002 m.

In biology (as in other sciences), the units used for measurement of length are based on the metre (m). Smaller units are obtained by dividing by 1000. For example:

- 1 millimetre (mm) = 1/1000 m
- 1 micrometre (μm) = 1/1000 mm (1/1000 000 m)

Larger units are obtained by multiplying by 1000. For example:

- 1 kilometre (km) = 1000 m

Another derived unit that is used frequently is the centimetre (cm). One centimetre is 1/100 of a metre, or 10 millimetres.

How big can we make things appear and how clearly can we see them?

Magnification describes any process that makes an object appear larger than it is. The early microscopes used by Hooke did not magnify greatly. Today, a photograph of an image produced by such a microscope could be further magnified by placing it on a photocopier and selecting 'enlarge'. However, this would merely produce a bigger version of a rather blurred image. There would not be any more detail.

This is a matter of **resolution**. Resolution is the ability of an instrument to distinguish between two points that are close together. If they cannot be resolved, they will be seen as one point and the detail seen will be limited.

How do optical microscopes work?

Optical microscopes have come a long way since the days of Hooke. The best ones can now produce clear, detailed images magnified up to 1250 times. The images can be viewed directly or, in some cases, through a computer-aided projection system on a screen. Unless your biology department is very well-funded, the microscopes that you use will not be quite of this standard! Nonetheless, standard microscopes used at A-level can produce clear images with magnification up to 400 times.

You need to have an idea of the size of organisms and cells. You may be asked to calculate the size of, for example, a bacterium from a drawing or photograph. It is easy to press the wrong button on a calculator and produce an answer that suggests a bacterium is 50 m long (it is now bigger than a whale!), rather than 5 μm long! If you understand the range of sizes of organisms, a moment's thought will tell you that the first answer is wrong.

(a) A modern optical research microscope with a computer-aided projection system
(b) A standard microscope used at A-level

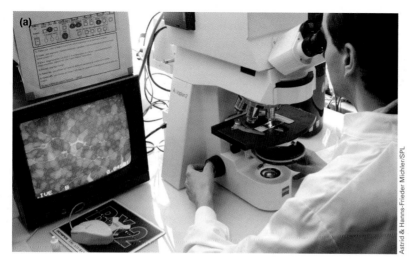

Astrid & Hanns-Frieder Michler/SPL

AJ Photo/SPL

An optical microscope passes rays of light through a specimen. As the light passes through, different parts of the specimen absorb different amounts and different wavelengths of light. Therefore, different shades and intensities of light pass through the two lenses to the eye from different parts of the slide. Regions that have absorbed more light appear darker because less light passes through. We say that they are **optically dense** regions.

Each lens refracts the light and, as a result, magnifies the image. The overall magnification of the microscope is the product of the magnification of the objective and eyepiece lenses. For example, an eyepiece with a magnification of 10 times and an objective with a magnification of 40 times produce an overall magnification of 400 (10 × 40).

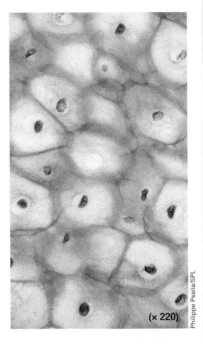

(× 220)

Philippe Psaila/SPL

The nucleus of each cell is more optically dense than the cytoplasm

However, no matter how good the lenses, the resolution of an optical microscope is limited by the wavelength of light itself. The best optical microscopes have a resolving power of 0.2 μm. This means that anything smaller than 0.2 μm is not visible as a separate object using an optical microscope.

Which part of the specimen is being observed?

When using an optical microscope, depth of focus is an issue. Remember, specimens may be up to one cell thick. The organelles are not distributed uniformly within cells so, by altering the depth of focus slightly, you can alter the appearance of the image.

Box 2.1 Preparing slides for light microscopy

To prepare a specimen for viewing under a light microscope, the specimen is placed on a glass slide. It can be stained to reveal subcellular structures, such as nuclei or starch grains. To view detail of cells under high magnification, the specimen should be no more than one cell thick (thinner if possible) to allow maximum transmission of light. You have probably prepared a slide of onion epidermis, which is a convenient way of obtaining plant tissue that is only one cell thick.

Specimens on commercially prepared slides (particularly of animal tissue) have frequently been:

- dehydrated
- stained
- embedded in wax
- sectioned (sliced) using a microtome (a device that produces slices of uniform thickness)

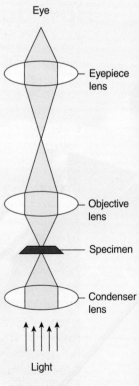

Figure 2.2 How an optical microscope produces an image

What are electron microscopes?

An electron microscope uses a beam of electrons, rather than light. Electrons have a shorter wavelength than light, so these microscopes have a higher resolving power than optical microscopes. This higher resolving power has allowed biologists to observe the ultrastructure of cells in much more detail than was possible with light microscopes.

There are two main kinds of electron microscope:

- the **transmission electron microscope** (TEM) passes a beam of electrons through a specimen (rather like a beam of light passing through a specimen in an optical microscope)
- the **scanning electron microscope** (SEM) directs a beam of electrons at a specimen and creates an image from the electrons that are *reflected* from the surface, rather than from those that pass through

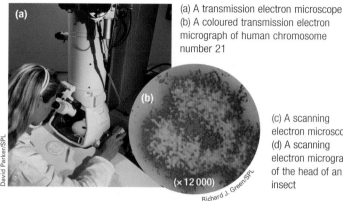

(a) A transmission electron microscope
(b) A coloured transmission electron micrograph of human chromosome number 21

(c) A scanning electron microscope
(d) A scanning electron micrograph of the head of an insect

David Parker/SPL

Richard J. Green/SPL

(×12 000)

Steve Gschmeissner/SPL

(×100)

Chris Taylor/CSIRO/SPL

Today's transmission electron microscopes can magnify up to 500 000 times and have a resolving power of 0.001 µm. This can reveal the internal structure of cell organelles and even produce outlines of large molecules, such as DNA.

A specimen for examination using a transmission electron microscope must be prepared in such a way that electrons can pass through it. Electrons are much more easily blocked than light, so the specimen must be exceptionally thin. One-cell thick would be much too thick because very few electrons would pass through. The specimen must also be examined in a vacuum as atoms and molecules in the air would affect the transmission of electrons. Preparation involves using chemicals (often the salts of heavy metals) to fix the specimen. However, there are negative consequences that may arise. For example, the preparation techniques may:

- alter the specimen from its original condition
- introduce 'artefacts' — structures not originally present

Figure 2.3
How a transmission electron microscope produces an image

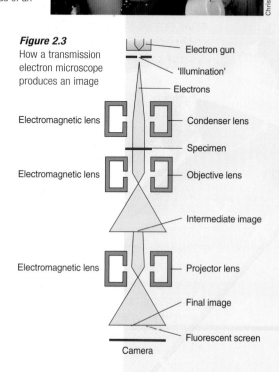

- Electron gun
- 'Illumination'
- Electrons
- Electromagnetic lens — Condenser lens
- Specimen
- Electromagnetic lens — Objective lens
- Intermediate image
- Electromagnetic lens — Projector lens
- Final image
- Fluorescent screen
- Camera

Despite this, it is believed that the images produced by transmission electron microscopes do show the true internal structure of cells and organelles. The images are consistent and are supported by other evidence. Therefore, it is reasonable to conclude that they are accurate.

The scanning electron microscope does not have quite the resolving power or magnification of a transmission electron microscope, but it provides a different view of a specimen. It forms an image from the electrons that are reflected from the specimen, so it produces a three-dimensional effect image of the surface of the specimen — for example, of an insect's head or of the inside surface of the nuclear membrane.

The specimen must be chemically treated, dried and coated with a thin film of gold before being viewed in a vacuum. This treatment may introduce artefacts or modify the appearance of the specimen. However, consistency of the images suggests that they are truly representative of the structures being viewed.

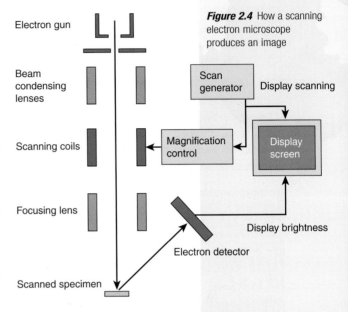

Figure 2.4 How a scanning electron microscope produces an image

How should we interpret micrographs?

Transmission electron micrographs are photographs taken through a transmission electron microscope.

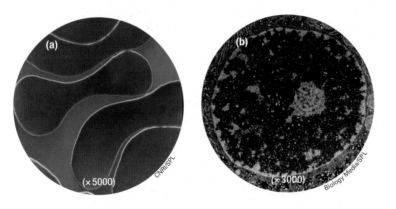

(a) Electron micrograph of red blood cells
(b) Electron micrograph of the nucleus of a liver cell

◄ A red blood cell consists almost entirely of haemoglobin. Therefore, it has uniform **electron density** and uniform brightness. The various structures in the nucleus of the liver cell have different compositions and electron densities, so there are different shades of light and dark.

Different shades

Electron-dense material allows few electrons to pass through. As a result, these types of material appear dark when viewed on the fluorescent screen of a transmission electron microscope. If a micrograph is of uniform brightness/darkness, the same number of electrons must be passing through all regions of the section. All regions, therefore, have the same electron density and are likely to have the same composition.

Different shapes

The appearance of a specimen under a microscope depends on a number of factors. First, from what angle is the specimen being viewed? Second, are we viewing the whole specimen? Red blood cells are often described as 'biconcave discs'. When viewed through an optical microscope, they usually appear round and are pale in the centre. When seen from the side, the biconcave aspect of their shape is apparent — they are thinner in the middle, which accounts for the 'paleness' here, when viewed from the front.

◀ Red blood cells are disc-shaped when seen from the front and biconcave when seen from the side.

However, red blood cells examined by a transmission electron microscope do not appear as biconcave discs. Why is this? When specimens are prepared for a transmission electron microscope, very thin sections are cut. Many sections are taken through a single cell. Different sections will have different shapes, depending on the angle at which each was cut. In addition, the red blood cells may be squashed or bent during the preparation, further distorting their shape.

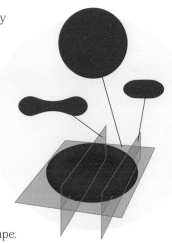

◀ Distortion can occur when red blood cells are squeezed through tiny blood vessels.

Figure 2.5 Sections can have different shapes, depending on the angle at which each was cut

Red blood cells viewed through an electron microscope

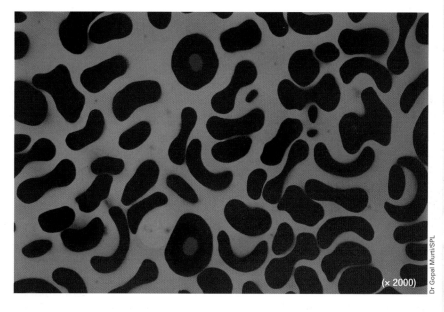

(× 2000)

Dr Gopal Murti/SPL

How big is it really?

If the size of the image on the micrograph (the apparent size) and the magnification are known, then we can calculate the actual size of the specimen. If a photograph of an object (the apparent size) is ten times bigger than the object (the actual size), then the object must be one-tenth the size of its picture. So, if the length in the photograph is 5 cm, the actual length of the object must be one-tenth of this — 0.5 cm. The equation relating actual size, apparent size and magnification can be written in three ways:

$$actual\ size = \frac{apparent\ size}{magnification}$$

$$magnification = \frac{apparent\ size}{actual\ size}$$

$$apparent\ size = actual\ size \times magnification$$

The apparent size and actual size must be in the *same units*.

Worked example

The diameter of a cell in a micrograph is 3.5 cm. The magnification is 450. What is the actual size of the cell in micrometres (μm)?

Answer

Information given:

- apparent size (diameter of cell in micrograph)
- magnification

The relevant form of the equation is:

$$actual\ size = \frac{apparent\ size}{magnification}$$

The answer is required in micrometres. Therefore, the apparent size (cm) must first be converted to micrometres:

1 cm = 10 mm; 1 mm = 1000 μm, so 1 cm = 10 000 μm
3.5 cm = (3.5 × 10 000) μm = 35 000 μm

The apparent size is 35 000 μm. Substituting into the equation:

$$\frac{apparent\ size}{magnification} = actual\ size$$

$$\frac{35\ 000}{450} = 77.8\ μm$$

The diameter of the cell is 77.8 μm.

In a question, the apparent size and actual size may be given in different units. In this case, you must convert the units of one of the measurements into the same units as the other.

How good are the different microscopes?

Optical microscopes, transmission electron microscopes and scanning electron microscopes have particular uses and limitations. For example, to see how cells are organised in a tissue or organ, it is better to use an optical microscope. The magnification and resolution are not as great as with a transmission electron microscope, but the field of view is wider and the cells can be viewed alive. However, to study the ultrastructure of cell organelles, a transmission electron microscope is needed — an optical microscope does not have the necessary magnification or resolution. A scanning electron microscope gives a highly resolved view of the surface of a specimen. The properties of optical microscopes and electron microscopes are compared in Table 2.1.

Table 2.1

Property	Optical microscope	Transmission electron microscope	Scanning electron microscope
Magnification	Maximum 1250 times	Maximum 500 000 times	Maximum 250 000 times
Focusing	Glass lenses: eyepiece (ocular lens) and objective lens	Electromagnetic lenses	Electromagnetic lenses
Resolution	Maximum 0.2 μm	Maximum 1 nm	Maximum 1 nm
Specimen preparation	Can be mounted in water or aqueous solution; can be alive; may be stained	Specimen is fixed with salts of heavy metals and viewed in a vacuum	Specimen is chemically treated, coated with a thin film of gold and viewed in a vacuum
Image	Viewed directly through the eyepiece	Viewed on a fluorescent screen	Viewed on a fluorescent screen
Limitations	Low resolution does not allow much subcellular detail to be observed	Specimens are always dead and may contain artefacts as a result of preparation techniques	Specimens are always dead and may contain artefacts as a result of preparation techniques
Advantages	Can be used to view live whole specimens under low magnification	Can produce high-magnification, high-resolution images of cells and organelles	Creates a 3-D effect image of the surface of the specimen

Ultrastructure of cells

There are two main types of cell — **prokaryotic** and **eukaryotic**. Prokaryotic cells are the most primitive. Most biologists believe that they resemble the first cells formed. Bacterial cells, including those of blue-green bacteria, are prokaryotic cells. Eukaryotic cells are more complex and contain more organelles. The cells of plants, animals, protoctistans and fungi are eukaryotic.

◄ Organelles are subcellular structures with a specific function.

Transmission electron micrograph of columnar epithelial cells

What's in an animal cell?

Not all animal cells contain all the same organelles and structures. Many cell types are specialised for a particular function, which means that certain organelles are more important than others. One type of cell that contains most of the organelles is the epithelial cell from the lining of the small intestine.

Figure 2.6
The ultrastructure of an epithelial cell from the small intestine

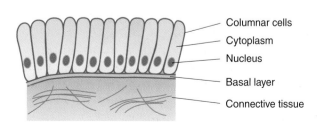

Columnar cells
Cytoplasm
Nucleus
Basal layer
Connective tissue

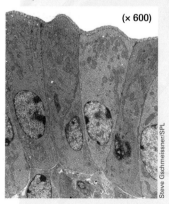

(× 600)

Steve Gschmeissner/SPL

Nucleus

The **nucleus** typically occupies about 10% of the volume of a cell. It has several components:

- The **nuclear envelope** is a double membrane that surrounds the nucleus. There are many **nuclear pores**, which allow the passage of some molecules between the nucleus and the cytoplasm.
- The **nucleolus** is an organelle within the nucleus. It is not membrane-bound. Its function is to synthesise the components of ribosomes, which then pass through the nuclear pores into the cytoplasm.

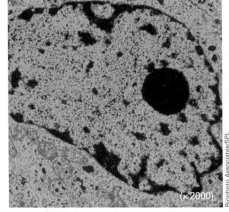

(× 2000)

Nucleotides (the building blocks of DNA and RNA) and some proteins can pass from the cytoplasm into the nucleus. Messenger RNA molecules pass from the nucleus into the cytoplasm. DNA molecules are too large to pass through the nuclear pores.

Electron micrograph of a nucleus

- **Chromatin** consists of DNA molecules bound with proteins called histones. For most of the cell cycle, the chromatin fibres are loosely dispersed throughout the nucleus. Just before a cell is about to divide, the chromatin condenses into distinct, recognisable structures called **chromosomes**.

Mitochondrion

Mitochondria are the sites of most of the reactions of aerobic respiration. They are surrounded by two membranes. The inner membrane is folded into **cristae** to increase the available surface area.

Some of the reactions of aerobic respiration take place in the fluid matrix. The folded inner membrane provides a large surface area for the electron-transport system, which produces most of the ATP.

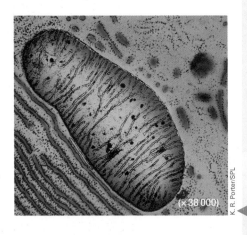

(× 38 000)

Transmission electron micrograph of a mitochondrion

ATP is the 'energy storage molecule' of cells. Energy released in respiration is stored in ATP molecules, to be released and used when needed. Cells that are very active (such as muscle cells or epithelial cells that absorb molecules from the gut) use a great deal of ATP and, therefore, contain many mitochondria.

Figure 2.7 Structure of a mitochondrion

Inside is the matrix, which contains Krebs cycle enzymes

Bounded by a double phospholipid bilayer membrane

The inner phospholipid bilayer is folded to form cristae

Ribosomes

Ribosomes are the sites of protein synthesis. They can be found free in the cytoplasm, but are also bound to the membrane system of the **endoplasmic reticulum**, forming rough endoplasmic reticulum. Each ribosome comprises two subunits that are made from ribosomal RNA and protein. The subunits are manufactured in the nucleolus. They leave the nucleus through nuclear pores and combine in the cytoplasm.

Endoplasmic reticulum

Endoplasmic reticulum (ER) is a membrane system found throughout the cytoplasm of eukaryotic cells. There are two types of endoplasmic reticulum:

- **Rough ER** has ribosomes on its surface and is responsible for the manufacture and transport of proteins. Protein molecules manufactured by the ribosomes pass through small pores into the lumen of the ER. They are then moved in a vesicle to the Golgi body.
- **Smooth ER** has no ribosomes on its surface. It is concerned with the synthesis of lipids. It is also associated with carbohydrate metabolism and detoxification.

◀ Rough ER is extensive in cells that manufacture a lot of protein, such as cells that manufacture enzymes to be secreted into the lumen of the intestine.

◀ Smooth ER is particularly extensive in brain and liver cells.

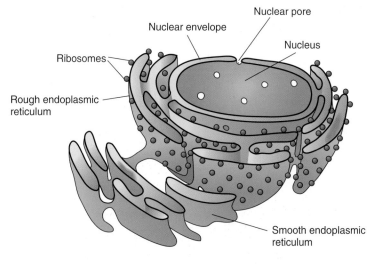

Nuclear pore

Nuclear envelope

Nucleus

Ribosomes

Rough endoplasmic reticulum

Smooth endoplasmic reticulum

Figure 2.8 Rough ER and smooth ER are continuous with each other and with the outer membrane of the nuclear envelope

Golgi apparatus

The **Golgi apparatus** consists of a number of flattened membrane-bound sacs in which proteins, and other molecules, are modified. For example, proteins may be converted into glycoproteins. Many of the modifications added in the Golgi apparatus act as a kind of 'tag' that determines the final destination of the molecule. Think of the Golgi apparatus as a cellular Post Office that labels and then distributes molecules!

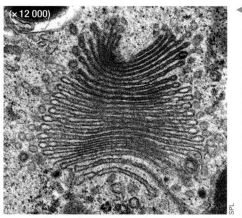

(×12 000)

SPL

◀ The Golgi apparatus is most extensive in secretory cells, such as cells that secrete digestive enzymes, nerve cells (secrete neurotransmitters) and B-cells (secrete antibodies).

The Golgi apparatus

Many of the modified molecules are released from the Golgi apparatus in vesicles that may be carried:

- to other parts of the cell
- to the plasma membrane, to pass out of the cell by exocytosis for use elsewhere

Some of these vesicles form the lysosomes.

Lysosomes

Lysosomes have no specialised internal structure and are surrounded by a single membrane. They are formed in the Golgi apparatus and contain digestive enzymes (produced in the rough ER) that break down cellular waste and debris.

Plasma membrane

The **plasma membrane** is the boundary membrane of all eukaryotic cells. It controls what passes into and out of the cell. There have been several models of membrane structure. The model currently thought to be the best representation is the **fluid mosaic model**. The basic structure is a **phospholipid bilayer** — a double layer of phospholipid molecules.

Different types of protein molecule are embedded in this bilayer:

- **ion-channel proteins** allow certain ions to pass through
- **transport proteins** carry molecules into or out of the cell
- **glycoproteins** act as cellular markers to allow the cell to be recognised

◄ Lysosomes are particularly abundant in phagocytic white blood cells.
Here, enzymes from the lysosomes digest foreign cells (and other materials) that have been engulfed.

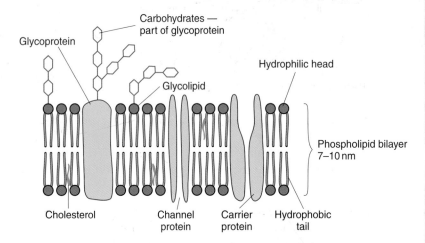

Figure 2.9 The fluid mosaic model of plasma membrane structure

The structure of the plasma membrane and the role of the various proteins in transporting substances in and out of the cell are discussed in more detail in Chapter 4.

In some cells, the plasma membrane is folded into structures called **microvilli**. These greatly increase the surface area of the cell; improving its ability to absorb substances.

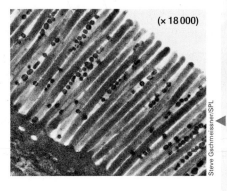

(× 18 000)

Steve Gschmeissner/SPL

Microvilli occur on the plasma membranes of cells that continuously absorb large amounts of substances. Examples include the epithelial cells that line the small intestine and those that ◄ line nephrons in the kidney.

False-colour transmission electron micrograph of microvilli on the lining of the gut

What's in a bacterial cell?

The prokaryotic cells of bacteria differ from eukaryotic cells in a number of ways:

- The DNA is not associated with protein to form true chromosomes. It is often referred to as naked DNA.
- The DNA is *circular*. The two 'ends' of the double helix of DNA are linked.
- Prokaryotes contain **plasmids**. These are small, circular pieces of DNA that can carry genes that confer resistance to antibiotics.
- The ribosomes are smaller.
- The cell wall is made from **peptidoglycan**. (Plant cell walls are made from cellulose; fungal cell walls from chitin.)
- There may be a **capsule** outside the cell wall. This prevents desiccation and provides some protection from digestion by enzymes in the gut of animals.
- They may possess **flagella** (singular flagellum).
- They may have small projections from the outer cell surface, called **pili** (singular pilus).
- They do *not* contain membrane-bound organelles such as mitochondria, chloroplasts and lysosomes.

> The bacterium that causes cholera has a capsule and a flagellum. The bacterium that causes pulmonary tuberculosis has neither ◀ of these structures.

> ✎ If you are asked to describe the features of a prokaryotic cell, list those that are present (peptido-glycan cell wall, small ribosomes, circular DNA) rather than those that are absent (e.g. membrane-bound organelles, nuclear envelope). If something is not there, how can it be a feature? However, if you are asked for *differences*, you can include structures that are absent.

Figure 2.10
Generalised diagram of a bacterial cell

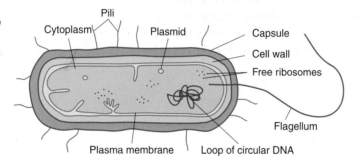

Pili — Cytoplasm — Plasmid — Capsule — Cell wall — Free ribosomes — Flagellum — Loop of circular DNA — Plasma membrane

Cell fractionation

Although much of the ultrastructure of organelles has been discovered from micrographs of whole cells, understanding the function of the organelles requires experimental study.

The organelles can be separated by **cell fractionation**. The technique is based on the fact that the masses of different organelles vary and depend on their size. When a mixture of organelles is spun in an ultracentrifuge, the various types settle out at different speeds. The large nucleus requires a relatively low centrifuge speed to make it settle out; the much smaller ribosomes require a much higher speed.

The technique, illustrated in Figure 2.11, is carried out as follows:

- The cell sample is stored in a suspension that is:
  - buffered — the neutral pH prevents damage to the structure of proteins, including enzymes
  - isotonic (of equal water potential) — this prevents osmotic water gain or loss by the organelles; gaining too much water could rupture the organelles
  - cool — this reduces the overall activity of enzymes released later in the procedure

- The cells are homogenised in a blender and filtered to remove debris.
- The homogenised sample is placed in an ultracentrifuge and spun at low speed — nuclei settle out, forming a pellet.
- The supernatant (the suspension containing the remaining organelles) is spun at a higher speed — chloroplasts settle out (if plant tissue is used).
- The supernatant is spun at a higher speed still — mitochondria and lysosomes settle out.
- The supernatant is spun at an even higher speed — the ribosomes, membranes and Golgi apparatus settle out.

◀ An ultracentrifuge can 'spin' a suspension at much higher speeds than a conventional centrifuge.

The ribosomes, membranes and Golgi complexes can be separated by another technique, called **density-gradient centrifugation**.

> *e* You may be asked to explain the reason for storing the cell sample in an isotonic solution. Candidates frequently make the mistake of suggesting that this will prevent osmotic damage to the *cells*. It is the *organelles* that are protected from osmotic damage. The cells themselves are burst open in the blender.

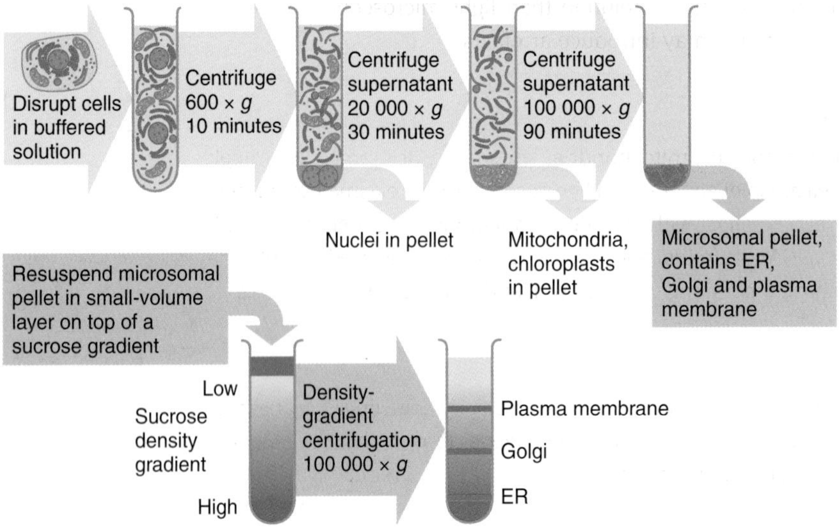

Disrupt cells in buffered solution

Centrifuge 600 × *g* 10 minutes

Centrifuge supernatant 20 000 × *g* 30 minutes

Centrifuge supernatant 100 000 × *g* 90 minutes

Nuclei in pellet

Mitochondria, chloroplasts in pellet

Microsomal pellet, contains ER, Golgi and plasma membrane

Resuspend microsomal pellet in small-volume layer on top of a sucrose gradient

Sucrose density gradient

Low

High

Density-gradient centrifugation 100 000 × *g*

Plasma membrane

Golgi

ER

Figure 2.11 How differential centrifugation separates cell components

Summary

Cell theory

- A cell is the smallest unit of life capable of independent existence.

Scale, magnification and resolution

- Dimensions are measured in metres or units derived from the metre, including:
  - kilometre, km = 1000 m
  - millimetre, mm = 0.001 m
  - micrometre, μm = 0.000 001 m (0.001 mm)

- Magnification produces an enlarged image of an object.
- Resolution is the ability to distinguish between two points that are close together.
- magnification $= \dfrac{\text{apparent size}}{\text{real size}}$

Apparent and real size must be measured in the *same units*.

Microscopes

- An optical microscope passes a beam of light through the specimen, which may be alive or dead, a small whole organism or a section.
- A transmission electron microscope passes a beam of electrons through a specimen, which must be a very thin section of either a cell or a small organism.
- A scanning electron microscope forms an image from electrons reflected from the surface of a specimen.
- Optically and electron-dense regions of a specimen allow little light or few electrons to pass through, respectively. Therefore, they appear as dark regions in the image.
- Transmission electron microscopes and scanning electron microscopes have higher magnification and better resolution than light microscopes. However, preparation of the specimen may introduce artefacts.

Cell ultrastructure

- Animal cells contain a nucleus, mitochondria, lysosomes, ribosomes, ER (rough and smooth) as well as Golgi apparatus, all enclosed within a plasma membrane.
- Prokaryotic cells are smaller, with no membrane-bound organelles but with naked and circular DNA (plasmids) and smaller ribosomes enclosed in a plasma membrane. They have a peptidoglycan cell wall and may also have a capsule.

Cell fractionation

- Cell fractionation separates the components of a cell by centrifugation, heavier organelles being isolated at lower centrifuge speeds. Prior to homogenisation followed by centrifugation, the tissue is refrigerated (to reduce the rate of metabolic reactions) in an isotonic solution (to prevent osmotic damage to the organelles following homogenisation) which is pH buffered (to prevent damage to the tertiary structure of cell proteins, including enzymes).

Questions

Multiple-choice

1 To convert millimetres to micrometres:
   A multiply by 1000
   B divide by 100
   C divide by 1000
   D multiply by 100

2 The resolving power of a microscope is its ability to:
   A produce an enlarged image
   B separate two nearby points
   C show a large field of view
   D reduce the depth of view
3 Actual size, apparent size and magnification are related by the formula:
   A actual size = apparent size × magnification
   B apparent size = $\dfrac{\text{actual size}}{\text{magnification}}$

   C magnification = apparent size × actual size
   D apparent size = actual size × magnification
4 In comparison with optical microscopes, transmission electron microscopes have:
   A higher resolution but lower magnification
   B lower resolution but higher magnification
   C lower resolution and lower magnification
   D higher resolution and higher magnification
5 In the scanning electron microscope, the image is created from:
   A electrons that pass through the specimen
   B electrons that are reflected from the specimen
   C both A and B
   D neither A nor B
6 The functions of the rough endoplasmic reticulum and the Golgi apparatus are related because:
   A proteins synthesised by the rough endoplasmic reticulum are modified by the Golgi apparatus
   B proteins synthesised by the Golgi apparatus are modified by the rough endoplasmic reticulum
   C lipids synthesised by the Golgi apparatus are modified by the rough endoplasmic reticulum
   D lipids synthesised by the rough endoplasmic reticulum are modified by the Golgi apparatus
7 In cell fractionation, the purpose of keeping the tissue sample in an isotonic solution in a refrigerator prior to homogenisation is:
   A to prevent osmotic damage to the cells and to reduce the metabolic activity of the cells
   B to prevent osmotic damage to the cells and to increase the metabolic activity of the cells
   C to prevent osmotic damage to the organelles and to reduce the metabolic activity of the cells
   D to prevent osmotic damage to the organelles and to increase the metabolic activity of the cells
8 Mitochondria have:
   A a double membrane surrounding the organelle
   B a fluid substance inside the organelle
   C both **A** and **B**
   D neither **A** nor **B**

9 The plasma membrane provides:

   A structural support for the cell and regulates which substances enter and leave

   B structural support for the cell but does not regulate which substances enter and leave

   C no structural support for the cell and does not regulate which substances enter and leave

   D no structural support for the cell but regulates which substances enter and leave

10 The principle behind separating cell organelles by ultracentrifugation is that:

   A the various organelles have different masses

   B the various organelles have different volumes

   C the various organelles have different shapes

   D the various organelles have different widths

Examination-style

1 (a) Copy and complete the table. Use a tick or cross to show if the feature is present or absent. *(3 marks)*

Type of cell	Feature		
	Chromosome	Cell wall	Mitochondrion
Red blood cell			
Epithelial cell from small intestine			
B-cell			

(b) A red blood cell is viewed under an optical microscope with a magnification of 750 times. It appears round but paler in the centre than at the edges. The diameter of the red blood cell is measured as 6 mm.

   (i) Explain the difference in colour between the edge and the centre of the cell. *(1 mark)*

   (ii) Calculate the actual diameter of the cell. Express your answer in micrometres (µm). *(2 marks)*

Total: 6 marks

2 The drawing is from a photograph taken through a transmission electron microscope. It shows a cell organelle.

(× 6625)

(a) Explain the evidence that shows that the photograph was not taken through:

   (i) an optical microscope *(1 mark)*

   (ii) a scanning electron microscope *(1 mark)*

(b) Calculate the length of the organelle. Express your answer in micrometres (µm). *(2 marks)*

(c) Describe the function of the organelle. *(3 marks)*

Total: 7 marks

3 The drawing below is taken from a transmission electron micrograph of a cell from the pancreas.

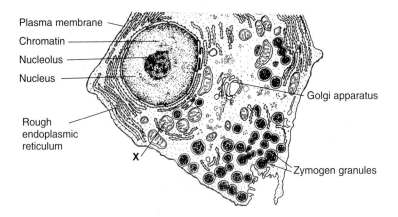

Plasma membrane
Chromatin
Nucleolus
Nucleus
Rough endoplasmic reticulum
X
Golgi apparatus
Zymogen granules

(a) Name the organelle labelled X. *(1 mark)*
(b) Explain why the nucleolus is not a uniform colour. *(2 marks)*
(c) The cell contains many zymogen granules, each of which contains inactive enzymes (which are proteins). Explain how this cell is able to:
  (i) synthesise large amounts of protein *(2 marks)*
  (ii) modify the protein into an enzyme and package it in zymogen granules *(2 marks)*
  Total: 7 marks

4 The following procedure was used to obtain different organelles from a sample of animal cells:

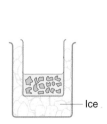  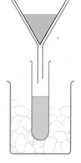 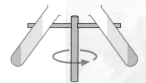

Tissue cut into small pieces and placed in a pH-buffered isotonic solution

Ice

Tissue put into blender/homogeniser

Mixture filtered to remove debris

Filtrate spun in centrifuge

(a) Explain why the tissue sample was:
  (i) placed in a pH-buffered solution that was isotonic with the tissue *(3 marks)*
  (ii) homogenised in a blender *(1 mark)*
(b) The organelles separated from the sample were: nucleus, mitochondria, Golgi apparatus, lysosomes, ribosomes and endoplasmic reticulum. Beginning with the first to settle out, list the organelles in the order in which they would have been separated by this process. *(1 mark)*

(c) There was an unusually large amount of Golgi apparatus in the sample. Suggest two possible origins for the tissue sample. Explain your answer. *(2 marks)*

Total: 7 marks

5 Acids in the stomach can damage stomach cells. Scientists investigating the effect of epidermal growth factor (EGF) on the extent of this damage carried out the following investigation.

Cells from a cell culture were washed in a buffer solution at pH 7 and then divided into four groups. Each group was placed in one of the following conditions for varying periods of time:

- pH 4.0 buffer solution only
- pH 4.0 buffer solution plus EGF (concentration 0.1 μg dm^{-3})
- pH 4.0 buffer solution plus EGF (concentration 1.0 μg dm^{-3})
- pH 4.0 buffer solution plus EGF (concentration 10.0 μg dm^{-3})

After 30 minutes, the cells were washed twice with normal (pH 7) buffer and the number of viable cells in each group was determined. Each investigation was repeated six times. The results are summarised on the graph below:

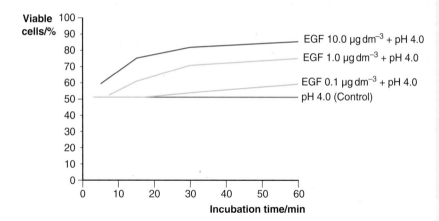

(a) Suggest why the cells were washed with normal buffer:
   (i) at the start of the investigation *(1 mark)*
   (ii) before the determination of viability was made *(2 marks)*
(b) Explain the importance of the investigation in which a pH 4 buffer only was used. *(2 marks)*
(c) Describe the pattern in the results with respect to:
   (i) incubation time *(2 marks)*
   (ii) concentration of EGF *(3 marks)*

In a separate investigation, the scientists investigated the effect of EGF on the pH of the intercellular medium (the medium between the cells). They used three treatments:

- cells maintained in a pH 7.0 buffer solution only
- cells maintained in a pH 4.0 buffer solution only
- cells maintained in a pH 4.0 buffer solution plus EGF (concentration 10 μg dm^{-3})

Their results are shown on the graph below.

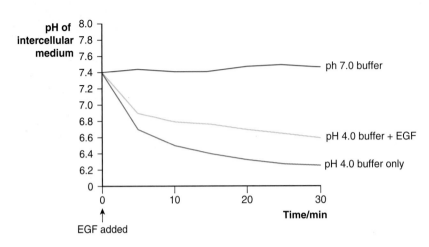

(d) (i) Suggest why the investigators chose to use the buffer
solutions and EGF concentration stated.   (*3 marks*)

(ii) Use the results to suggest how EGF might reduce acid
damage to cells.   (*2 marks*)

(**Total: 15 marks**)

Chapter 3

We must digest our food before we can use it

This chapter covers:
- the structure and functioning of the digestive system
- the digestion of starch
- enzymes as biological catalysts
- the way in which enzymes are able to catalyse reactions
- models of enzyme action:
  - the lock-and-key model
  - the induced-fit model
- factors affecting enzyme action:
  - temperature
  - pH
  - substrate concentration
  - enzyme concentration
  - inhibitors

Putting food into our mouths and swallowing does not get the food into our bodies. The food is in the first part of a lumen (space) that runs all the way through our body. This is easy to picture if you think of a worm.

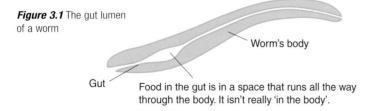

Figure 3.1 The gut lumen of a worm

Worm's body

Gut

Food in the gut is in a space that runs all the way through the body. It isn't really 'in the body'.

Before we can make use of the molecules in our food, we must get them out of the gut and into the bloodstream to be distributed around the body. However, many molecules, for example starch, lipids and proteins, are too large to be absorbed from the gut. They must be hydrolysed (broken down) into smaller molecules first. This process of converting food molecules into a form in which they can be absorbed is called **digestion**. In the case of starch, α-glucose is eventually formed. **Hydrolytic enzymes** control these reactions.

The hydrolysis of starch results from adding water to the α-1,4-glycosidic bonds between the glucose molecules that make up starch molecules. This reaction would take place only extremely slowly if it were not **catalysed** (made to go faster) by an enzyme, in this case amylase. All enzymes are **biological catalysts** (see pages 48–51).

What is the human digestive system like?

It is rather more complex than that of a worm, as Figure 3.2 shows.

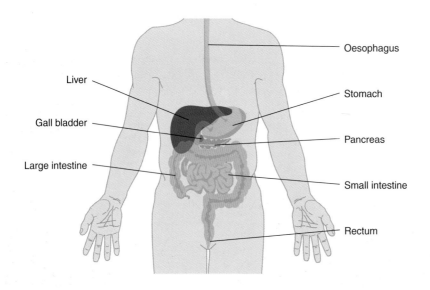

Oesophagus
Liver
Stomach
Gall bladder
Pancreas
Large intestine
Small intestine
Rectum

Figure 3.2 Food is ingested and chewed in the mouth (where some digestion of starch takes place). It is then swallowed and digested in the stomach, duodenum and ileum. The products of digestion are absorbed in the ileum and undigested material passes along the colon, where water and some salts are absorbed, before being egested in the faeces

What happens in the various parts of the digestive system?

Obviously, food is digested — end of chapter and turn the page. However, there is more to it than that. Food, in the form that we eat it, cannot be digested easily. It must be prepared for digestion and then it must be digested. The products of digestion are then absorbed and, after reclaiming as much as possible of the water that was mixed with the food, undigested materials pass out of the gut.

The digestion of pasta is used to illustrate this. The main nutrient in pasta is starch.

How is pasta prepared for digestion?

When we ingest pasta, we chew it and then swallow it. As we chew the pasta, saliva (from the salivary glands) is mixed with it. Saliva contains an enzyme called salivary amylase, which begins the digestion of starch. Chewing produces a ball of food called a **bolus**, which is then swallowed. The bolus is squeezed down the oesophagus by a process called **peristalsis**. In the stomach, the bolus is mixed and churned with gastric juice. This contains hydrochloric acid to kill bacteria that enter with the food, and an enzyme (**pepsin**) that begins the digestion of proteins.

After a period of churning the food with gastric juice, the state of the food (pasta) is changed from a solid to a milky liquid called **chyme**.

Saliva and gastric juice are digestive secretions (digestive juices) released by glands into the gut lumen. Salivary glands release saliva; glands in the stomach wall release gastric juice. Pancreatic juice (from the pancreas) and bile (produced in the liver and stored in the gall bladder) are also digestive secretions. All these secretions are mainly water, so a considerable amount of water is added to food as it moves along the gut.

How is the starch in pasta digested?

The chemical reactions of digestion proceed much faster in the liquid medium of chyme. Some digestion of starch takes place in the mouth, but this ceases in the stomach because the acidity of gastric juice denatures salivary amylase and stops it from working. However, when the chyme is released from the stomach into the duodenum, it is mixed with pancreatic juice, which contains another amylase — pancreatic amylase — that continues to hydrolyse starch. The action of amylase on starch produces molecules of the disaccharide, maltose.

To complete the digestion of starch, a second enzyme is needed. This enzyme is maltase. It catalyses the hydrolysis of maltose into α-glucose molecules. However, maltase is not secreted into the lumen of the gut like amylase is. Instead, it forms part of the plasma membrane of some of the epithelial cells that line the small intestine (both duodenum and ileum). These cells have plasma membranes that are folded into microvilli on the side in contact with the lumen of the gut.

(a) The inner wall of the small intestine has many villi
(b) Most epidermal cells on the villi have microvilli to increase their surface area for absorption

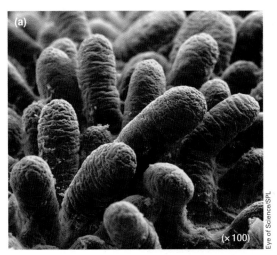

(a) (×100) Eye of Science/SPL

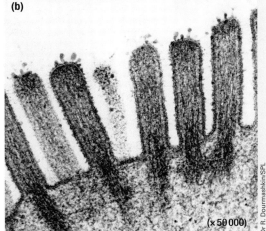

(b) (×50 000) Dr. R. Dourmashkin/SPL

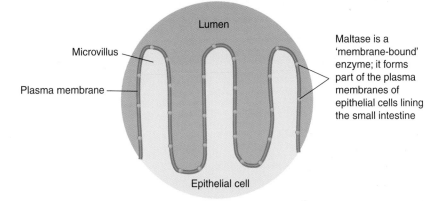

Lumen

Microvillus

Plasma membrane

Epithelial cell

Maltase is a 'membrane-bound' enzyme; it forms part of the plasma membranes of epithelial cells lining the small intestine

Figure 3.3 The location of maltase

Maltose molecules diffuse towards the enzyme and bind with it. Water molecules react with it to hydrolyse the α-1,4-glycosidic bond between the two glucose molecules.

The stages in the digestion of starch are summarised in Table 3.1.

Table 3.1 The digestion of starch

Stage in digestion	Enzymes involved	Notes on reaction
Starch to maltose	Amylase from salivary glands and pancreas.	• Amylase catalyses the hydrolysis of α-1,4-glycosidic bonds between α-glucose units in the starch molecule. • Optimum pH is neutral to slightly alkaline. • Reaction begins in the buccal cavity and recommences in the duodenum. • Acid pH of the stomach denatures salivary amylase.
Maltose to glucose	Maltase in the plasma membranes of the cells of the ileum.	• Enzymes bound to the plasma membrane of the microvilli hydrolyse • α-1,4-glycosidic bonds between the two α-glucose units in maltose molecules.

Box 3.1 Digestion of the protein in pasta

Protein digestion follows a similar pattern to the digestion of starch. In the stomach, the large protein molecules are broken down into smaller chains of amino acids. These are then digested into pairs of amino acids in the small intestine. The pairs are finally separated by membrane-bound enzymes in the plasma membranes of epithelial cells of the small intestine.

The products of digestion, including α-glucose from starch, are absorbed through the epithelium that lines the small intestine, into the bloodstream. Chapter 4 describes how this happens. Substances that are not absorbed pass into the colon (large intestine) where much of the water secreted in the digestive juices is reabsorbed. The material left forms the faeces, which are stored temporarily in the rectum before being egested through the anus.

◀ The enzyme lactase, which hydrolyses lactose into β-glucose and galactose, is another membrane-bound enzyme. In some people, this enzyme ceases to be made as they grow older, giving rise to **lactose intolerance** (see pages 5–6).

What are enzymes?

You should appreciate that not all enzymes are involved in digestion. In fact, digestive enzymes make up only a tiny proportion of the enzymes that are found in our bodies. However, all enzymes have some things in common. They catalyse the many biochemical reactions that take place inside cells and outside cells. This ensures that the reactions take place efficiently when the enzyme is present and hardly at all when the enzyme is absent.

Box 3.2 The discovery of enzymes

Much of the early work on enzymes was related to fermentation because wine producers were interested in ways of improving the process. The first significant discovery came in 1833 when Payen and Persoz produced an extract from malt (germinated cereal seeds) that hydrolysed sucrose into fructose and α-glucose. However, this was largely ignored because Louis Pasteur and other influential biologists believed that fermentation could be carried out only by intact, living microorganisms, such as yeast cells. However, more and more evidence was found showing that chemicals *in* the cells were catalysing these reactions, rather than whole cells. In 1860, Berthelow obtained an extract from yeast that also hydrolysed sucrose into fructose and α-glucose. The argument was finally sealed by Eduard Buchner who, in 1897, produced what he demonstrated to be a cell-free extract of yeast that carried out the complete fermentation of sugar. As more and more evidence accumulated, the 'whole-cell' model of fermentation was proved to be wrong. The chemical (enzymic) model took its place. This model has been extended to include all the reactions that occur in living cells, not just fermentation.

What are catalysts?

A **catalyst** is a substance that speeds up a reaction. The nature of the reaction is unaltered; only the speed at which it takes place is affected. There is no overall change to:

- the nature of the products
- the energy change that takes place during the reaction
- the catalyst itself

This is true of all catalysts, including enzymes, which are **biological catalysts**. Catalysis enables biochemical reactions inside cells to take place quickly, at a temperature that will not damage the structure of the cell. To understand how this is possible, we must consider how chemical reactions take place.

Imagine a reaction in which substance A reacts with substance B to form substance AB. We can write an equation for this as:

$$A + B \rightarrow AB$$

However, this does not tell the whole story. The equation gives only the **reactants** (starting materials) and the **products** — it does not show how the reaction takes place. Before the final products are formed, the reactant molecules collide and enter a **transition state**, in which bonds in the molecules become strained. The molecules are **activated**. In this state, there is more likelihood of strained bonds breaking and new bonds forming — in other words, that the reactant molecules will react to form the products. However, under normal conditions, very few reactant molecules have sufficient kinetic energy to enter the transition state on collision, so the reaction proceeds slowly. A catalyst effectively lowers the energy required (the **activation energy, E_a**) for the molecules to enter the transition state, by providing an alternative pathway for the reaction that has lower activation energy (Figure 3.4). More reactant molecules consequently have this lower energy requirement and so the reaction proceeds more quickly (Figure 3.5).

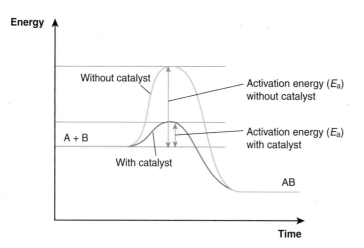

Figure 3.4 Effect of a catalyst on the activation energy of a reaction

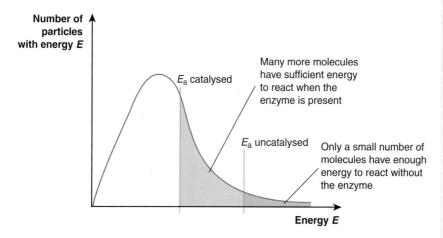

When reactant molecules bind to a catalyst, the binding strains the bonds and allows them to enter the transition state with a much lower energy input (lower activation energy) than would normally be the case. In this way, catalysts are able to speed up chemical reactions.

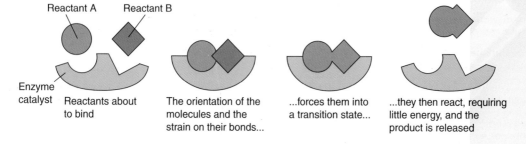

Most (but not all) biological catalysts are enzymes. Enzymes are globular proteins with a specific tertiary structure. The most significant part of this tertiary structure is the **active site**. This region of the enzyme molecule binds with **substrate** (reactant) molecules to form an **enzyme–substrate complex**. While bound to the active site, the reactants react and the product is released. The active site is shaped so that only certain substrate molecules are able to bind, and so each enzyme can catalyse only one reaction. The enzyme is said to be **specific**.

◀ Recently, it has been shown that some RNA molecules can catalyse certain biological reactions.

How do enzymes work?

There are two models of enzyme action — the **lock-and-key model**, first proposed in 1894 by a German biochemist named Fischer, and the **induced-fit model**, proposed in 1958 by Koshland.

The lock-and-key model

This model proposes that the shapes of the substrate molecules are *complementary* to that of the active site, rather like the shape of a key is complementary to that of the lock it fits.

Enzyme and substrate

A complex of enzyme and substrate allows reaction

Products are released and the enzyme is free to accept a new substrate molecule

Figure 3.7 The lock-and-key model of enzyme action

This model can explain enzyme specificity but does not explain how the substrate molecules become strained in order to enter the transition state.

> *℮* In an examination, be careful not to write, 'The shape of the substrate is the same as the active site'. It is not — if it were, they would not be able to bind. One egg cannot sit inside another egg because they are the same shape. However, an egg can sit in an eggcup because the shapes are complementary.

The induced-fit model

This model suggests that the binding of substrate molecules to the active site produces a conformational change (change in shape) in the active site and in other regions of the enzyme molecule. This conformational change puts the substrate molecules under tension, so they enter the transition state and are able to react.

As the enzyme and substrate bind, a change of shape occurs

The reaction proceeds as the enzyme and substrate bind

Products are released and the enzyme returns to its original shape

Figure 3.8 The induced-fit model of enzyme action

The induced-fit model is now generally accepted as offering a better explanation of enzyme action, but both models suggest that enzyme-controlled reactions proceed in two stages:

◀ Look back at Figure 3.6. It also shows the induced-fit model.

substrate + enzyme → enzyme–substrate complex

enzyme–substrate complex → enzyme + products

The rate of a chemical reaction is the rate at which reactants are converted into products. In the case of an enzyme-controlled reaction, this is determined by how many molecules of substrate bind with enzyme molecules to form enzyme–substrate complexes. The number of molecules of reactants that form enzyme–substrate complexes with each molecule of an enzyme, per second, is the **turnover rate** of the enzyme.

Box 3.3 Rate of reaction and turnover rate

Rate of reaction and turnover rate/enzyme activity are not the same. Rate of reaction describes the amount of substrate used or product formed irrespective of how many enzyme molecules catalyse the reaction. Enzyme activity is a measure of the fraction of maximum turnover at which each enzyme molecule is working. The rate of a reaction in which ten molecules of enzyme catalyse the reaction at maximum turnover will be slower than one in which 30 molecules of the same enzyme catalyse the same reaction at 70% maximum turnover. Suppose the turnover rate is 100 molecules per second. In the first example 1000 (100 × 10) molecules of substrate per second would form enzyme–substrate complexes. In the second example, 2100 (70 × 30) molecules of substrate per second would form enzyme–substrate complexes, increasing the rate of reaction even though the turnover rate/enzyme activity is lower.

What affects the efficiency of enzymes?

The turnover rate and, therefore, the activity of the enzyme is influenced by a number of external factors, including:
- temperature
- pH
- substrate concentration
- the presence of inhibitors

How hot must it be?

When the temperature is raised, particles in a system are given more kinetic energy, which has two main effects.
- 'Free' particles move around more quickly. This increases the probability that a substrate particle will collide with an enzyme molecule.
- Particles within a molecule vibrate more energetically. This puts strain on the bonds that hold the atoms in place. Bonds begin to break and, in the case of an enzyme, the shape of the molecule and the active site, in particular, begin to change. The enzyme begins to **denature**.

The activity of an enzyme at a given temperature is a balance between these two effects. If the raised temperature results in little denaturation but a greatly increased number of collisions, the activity of the enzyme will increase. If the increased temperature causes significant denaturation, then despite the extra collisions, the activity of the enzyme will probably decrease. The temperature at which the two effects just balance each other is the **optimum temperature** for that enzyme. Any further increase in temperature will cause increased denaturation that will outweigh the effects of extra collisions.

A decrease in temperature means that fewer collisions will occur.

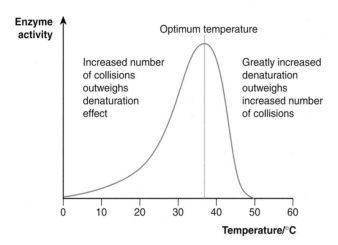

Figure 3.9 Effect of temperature on enzyme activity

Note that the graph is not symmetrical. Above the optimum temperature, the enzyme denatures very quickly to the point at which the shape of the active site has changed so much that an enzyme–substrate complex cannot form. At this point, the turnover rate is zero.

Box 3.4 Optimum temperature

Enzymes do not all have the same optimum temperature; they are adapted to work most efficiently within the organism in which they are found. For example, the optimum temperature for enzymes is:

- around 37°C (normal body temperature) in humans
- less than 5°C in plants growing in the Arctic
- over 90°C in bacteria that live in hot springs (thermophilic bacteria)

What pH do enzymes work best in?

pH is a measure of the hydrogen ion concentration of a solution or other liquid system. The pH scale ranges from 0 to 14. Solutions with a pH of less than 7 are acidic, those with a pH of more than 7 are alkaline and a solution with a pH of exactly 7 is neutral.

pH is an inverse logarithmic scale. Each pH unit represents a ten-fold change in hydrogen ion (H^+) concentration. pH 0 represents the highest H^+ concentration. A pH 1.0 solution has one-tenth (0.1) of this H^+ concentration; a pH 4 solution has one ten-thousandth (0.0001). pH 14 represents the lowest H^+ concentration.

The majority of enzymes in humans function most efficiently within the pH range 6.0–8.0, although the optimum pH of pepsin (an enzyme found in the stomach) is between pH 1.0 and pH 3.0. Significant changes in pH can affect an enzyme molecule by:

- breaking ionic bonds that hold the tertiary structure in place, leading to denaturation of the enzyme molecule
- altering the charge on some of the amino acids that form the active site, making it more difficult for substrate molecules to bind

These effects occur if the pH becomes either more acidic or more alkaline.

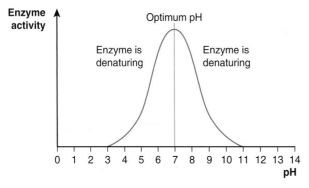

Figure 3.10 Effect of pH on enzyme activity

Does the concentration of the substrate matter?

The activity of an enzyme depends on the number of substrate molecules per second that bind to form enzyme–substrate complexes. This is linked to the number of substrate molecules present. A small number of substrate molecules means few collisions and the formation of few enzyme–substrate complexes. Increasing the concentration of the substrate means more collisions and more enzyme–substrate complexes. Since increasing the substrate concentration increases the activity of each enzyme molecule, the overall rate of reaction is increased. Eventually, a situation could be reached in which, because of the high substrate concentration, each enzyme molecule is working at maximum turnover — that is, each active site is binding with substrate molecules continually and there is no 'spare capacity' in the system. Increasing the substrate concentration beyond this point cannot increase the activity of the enzyme or the rate of reaction because all the active sites are occupied.

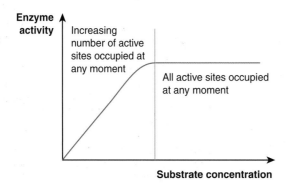

Figure 3.11 Effect of substrate concentration on enzyme activity when enzyme concentration is constant

Now think about an enzyme-controlled reaction, where the concentration of enzyme is kept constant. As the reaction proceeds, the substrate concentration decreases, because, as each molecule of substrate reacts, there will be fewer remaining. With fewer substrate molecules left, the number of collisions per second between enzyme and substrate decreases, so the reaction slows down. The turnover rate of each enzyme molecule decreases with time. Therefore, the reaction rate also decreases.

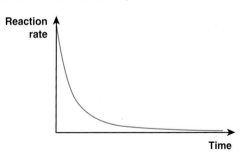

Figure 3.12 The change in reaction rate of an enzyme-controlled reaction over a period of time

How much enzyme should there be?

Assuming a constant large supply of substrate molecules, each enzyme molecule will work at maximum turnover. Therefore, the reaction rate will be directly proportional to the number of enzyme molecules — the concentration of the enzyme. Increasing the concentration will increase the reaction rate.

However, increasing the concentration of the enzyme will *not* increase the *activity* of the enzyme. Each enzyme molecule will be working at maximum turnover, so the activity of the enzyme is likely to remain constant.

How do inhibitors affect enzymes?

Inhibitors are substances that prevent enzymes from forming enzyme–substrate complexes and so stop, or slow down, catalysis of the reaction.

Competitive inhibitors

Competitive inhibitors have molecules with shapes that are complementary to all, or part, of the active site of an enzyme. They are often similar in shape to the substrate molecules. They can bind with the active site and prevent substrate molecules from binding. The binding is only temporary and the competitive inhibitor is quickly released.

The overall effect on the rate of reaction depends on the relative concentrations of substrate and inhibitor molecules. Each molecule of competitive inhibitor can inhibit (temporarily) one enzyme molecule — but only if it can collide with the enzyme molecule and bind with the active site. To do this, it must *compete* with the substrate molecules for the active site — hence the name, competitive inhibitor.

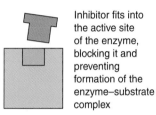

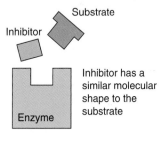

Figure 3.13 A competitive inhibitor blocks the active site so substrate molecules cannot bind

If there were 99 substrate molecules for every inhibitor molecule, then 99% of the collisions would be between enzyme and substrate and only 1% between enzyme and inhibitor. So, at any one time, only 1% of the enzyme molecules would be inhibited and the reaction would proceed at 99% of the maximum rate. If the ratio were 90 substrate molecules to ten inhibitor molecules, there would be 10% inhibition and the reaction rate would fall to 90% of maximum.

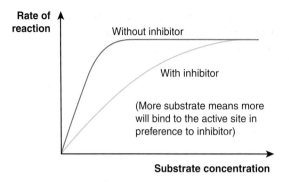

Figure 3.14 Effect of substrate concentration on inhibition by a competitive inhibitor

Non-competitive inhibitors

Non-competitive inhibitors do not compete for the active site. Instead, they bind to another part of the enzyme called the **allosteric site**. This produces a **conformational change** (change in shape) in the part of the enzyme molecule that includes the active site. As a result, the active site can no longer bind with the substrate and so cannot catalyse the reaction.

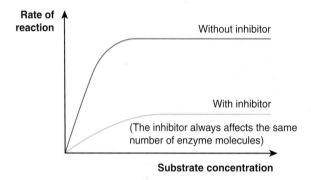

Figure 3.15 Effect of substrate concentration on inhibition by a non-competitive inhibitor

The effectiveness of a non-competitive inhibitor is not affected by the concentration of the substrate. Suppose there are enough inhibitor molecules to bind with the allosteric sites of 80% of the enzyme molecules. Eighty per cent of the enzyme molecules will be inhibited irrespective of the number of substrate molecules (as the substrate and inhibitor are not competing for the same site) and the reaction rate will drop to 20% of maximum.

Inhibitors and the regulation of cell metabolism

Many substances are produced in cells as a result of a series of reactions, which can be represented as:

$$A \xrightarrow{e_1} B \xrightarrow{e_2} C \xrightarrow{e_3} D$$

where e_1, e_2 and e_3 are enzymes catalysing the reactions.

All the reactions in this sequence are enzyme-controlled. Therefore, inhibition of any of these enzymes will interrupt the process. However, the main function is to produce substance D for use by the cell. If the requirement for substance D decreases, or ceases altogether, then the concentration of D will increase within the cell. This is at least inefficient (producing something that is not being used) and may be potentially harmful because high concentrations could be toxic. Such reaction sequences are often controlled by **end-product inhibition**. The end product (D) inhibits the enzyme controlling the first stage of the reaction sequence, thus:

$$e_1 \quad e_2 \quad e_3$$
$$A \rightarrow B \rightarrow C \rightarrow D$$

Substance D normally acts as a non-competitive inhibitor, which prevents the enzyme e_1 from catalysing the reaction that converts A to B. As a result, the entire reaction sequence is halted. When the inhibition is removed, the reaction sequence can start again. Sometimes, a decrease in the concentration of D is sufficient to remove the inhibition. In other cases, the inhibitor must be removed by a substance known as an **activator**.

Enzymes are often kept in an inactive state in a cell by an inhibitor and are activated by a specific activator only when they are required.

Summary

The digestive system

- The digestive system comprises the mouth, oesophagus, stomach, small intestine (duodenum and ileum), large intestine (colon), rectum and anus, as well as glands (salivary glands, gastric glands, pancreas and liver) that produce and secrete digestive juices into the system.
- Many food molecules need to be digested because they are too large to pass through the gut wall into the blood stream.
- Large food molecules are hydrolysed to smaller ones; the reactions are controlled by hydrolytic enzymes, which are biological catalysts.
- Starch is digested by amylase to maltose and then by maltase to α-glucose.
- Amylase is produced by the salivary glands and by the pancreas; salivary amylase acts in the mouth but is not active in the acidic stomach. Pancreatic amylase acts in the lumen of the gut. Maltase is a membrane-bound enzyme found in the plasma membranes of epithelial cells lining the small intestine.

Enzymes and catalysis

- A catalyst speeds up a chemical reaction with no effect on:
  - the products formed
  - the energy change of the reaction
  - the nature of the catalyst itself
- A catalyst speeds up a reaction by lowering the activation energy required for reactants to enter the transition state.
- Nearly all biological catalysts are enzymes. They are globular proteins with a specific tertiary shape, part of which forms an active site.
- A substrate molecule binds with the active site to form an enzyme–substrate complex. This then forms the products. The products are released from the enzyme molecule, which is unaltered.

- The lock-and-key model of enzyme action suggests a rigid structure for the enzyme molecule, with the shape of the substrate and active site being complementary to each other. This model explains enzyme specificity but not how the transition state is achieved.
- The induced-fit model of enzyme action suggests that binding of the substrate induces a conformational change in enzyme structure, which puts the substrate molecule under tension, causing it to enter the transition state.
- The number of substrate molecules that bind to the active site of an enzyme molecule per second is the turnover rate.

Factors affecting enzyme activity

- Temperature — below the optimum temperature, the low level of kinetic energy limits the number of enzyme–substrate complexes formed; above the optimum temperature, denaturation of the enzyme prevents binding of the substrate.
- pH — above and below the optimum pH, changes occur in the tertiary structure of the enzyme molecule and in the charges on the amino acids making up the active site; both prevent binding of the substrate.
- Substrate concentration — if the concentration of enzyme remains constant, increasing the substrate concentration increases the number of enzyme–substrate complexes formed until, at any one time, all the active sites are occupied; the rate of reaction increases to its maximum.
- Enzyme concentration — if the substrate concentration is high and constant, increasing the enzyme concentration increases the rate of reaction.
- Inhibitors:
  - competitive inhibitors have molecules that are often similar in shape to the substrate molecules and therefore compete for the active site on the enzyme; the extent of the inhibition depends on the ratio of substrate molecules to inhibitor molecules
  - non-competitive inhibitors bind to a region away from the active site, producing a conformational change in the enzyme that prevents the substrate from binding; the extent of the inhibition is independent of the substrate concentration

Questions

Multiple-choice

1 Digestion is necessary to:
  A ensure that large food molecules are converted to smaller ones
  B allow the effective absorption of food molecules
  C make the most effective use of the food we eat
  D all of the above

2 Enzymes are specific because:
  A each enzyme molecule has a unique tertiary structure
  B each has a specific active site
  C the shape of the substrate molecule is complementary to that of the active site
  D all of the above

3 The induced-fit model of enzyme action proposes that:
  A there is a conformational change in the enzyme molecule and the substrate molecule binds unaltered
  B there is no conformational change in the enzyme molecule and the substrate molecule binds unaltered
  C there is no conformational change in the enzyme molecule and the substrate molecule is put under tension as it binds
  D there is a conformational change in the enzyme molecule and the substrate molecule is put under tension as it binds

4 Maltase is:
  A a free enzyme that hydrolyses starch to maltose
  B a membrane-bound enzyme that hydrolyses starch to maltose
  C a membrane-bound enzyme that hydrolyses maltose to α-glucose
  D a free enzyme that hydrolyses maltose to α-glucose

5 The optimum temperature of an enzyme is the temperature at which:
  A most collisions between enzyme and substrate occur
  B the maximum number of enzyme–substrate complexes are formed
  C the enzyme is denatured
  D all the active sites are occupied at any given moment

6 Solutions with an extreme pH can:
  A alter the charge on amino acids in the enzyme's active site
  B reduce the number of enzyme–substrate complexes formed
  C alter the tertiary structure of the enzyme molecule
  D all of the above

7 Non-competitive inhibitors can:
  A bind with the active site to prevent substrate molecules from binding
  B change the shape of the active site to prevent substrate molecules from binding
  C alter the charge on amino acids in the active site to prevent substrate molecules from binding
  D all of the above

8 In an enzyme-catalysed reaction at constant enzyme concentration, increasing the concentration of substrate molecules will:
  A increase the rate of reaction to a maximum, then level off
  B increase the rate of reaction indefinitely
  C increase the rate of reaction to an optimum, then decrease the rate of reaction
  D have no effect initially, then increase the rate of reaction

9 Which of the following graphs best represents the effect of temperature on enzyme action?

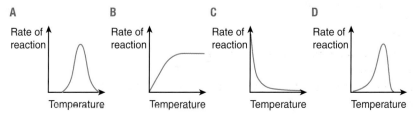

10 Most digestion of starch takes place in the:

   A mouth

   B stomach

   C pancreas

   D small intestine

Examination-style

1 **(a)** The diagram shows the main parts of the human digestive system.

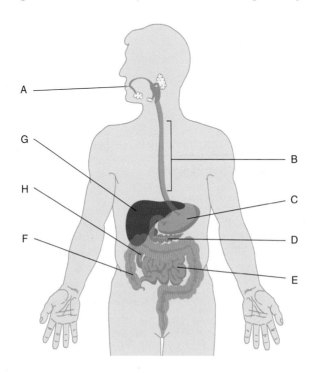

   (i) Name the structures labelled B, C and H. *(3 marks)*

   (ii) Give the letter of the structure that reabsorbs water secreted during digestion. *(1 mark)*

(b) The diagram shows how, during digestion, a molecule of maltose is converted into two molecules of α-glucose.

$$+ \quad H_2O$$

(i) Name the process taking place. (*1 mark*)

(ii) Explain the role of the enzyme maltase in this reaction. (*3 marks*)

Total: 8 marks

2 The graph below shows the effect of substrate concentration on the rate of an enzyme-controlled reaction under three conditions:

- a competitive inhibitor present
- a non-competitive inhibitor present
- no inhibitor present

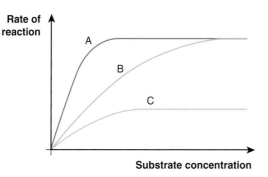

(a) Which of the three curves represents the effect of a competitive inhibitor? Explain your answer. (*3 marks*)

(b) Explain how a non-competitive inhibitor produces its effect. (*3 marks*)

Total: 6 marks

3 The rate of enzyme action depends on a number of factors, including the concentration of the substrate. The graph below shows the rate of reaction of a human enzyme at different substrate concentrations. The investigation was carried out at 25°C.

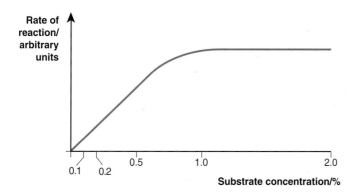

(a) Explain the shape of the graph in terms of kinetic theory and enzyme–substrate complex formation:

(i) from substrate concentration 0.1% to 0.5%

(ii) from substrate concentration 1.0% to 2.0% (*4 marks*)

(b) Sketch, on the graph, the curve you would expect if the experiment had been carried out at 35°C rather than 25°C. (*1 mark*)

(c) The graph below represents an energy level diagram of a reaction proceeding without an enzyme and the same reaction with an enzyme.

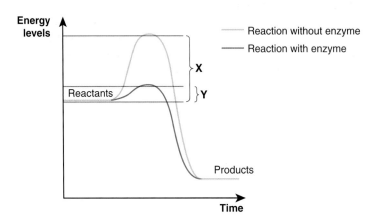

(i) Describe two ways in which the energetics of the two reactions are similar. *(2 marks)*

(ii) Explain the differences between the regions marked X and Y on the diagram. *(1 mark)*

Total: 8 marks

4 Hydrogen peroxide decomposes slowly to give water and oxygen. The reaction is catalysed in many cells by catalase. In an investigation into the effect of temperature on the rate of decomposition of hydrogen peroxide by catalase, the rate of reaction was measured by collecting the oxygen produced in a 10 minute period at different temperatures. All other factors were kept constant. The results are summarised in the graph below.

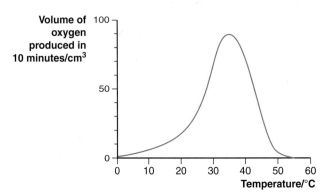

(a) (i) According to this graph, what is the optimum temperature of catalase? *(1 mark)*

(ii) Explain why this might not be an accurate estimate of the optimum temperature. *(2 marks)*

(b) In a control experiment (no enzyme present but all other factors the same as in the other experiments) carried out at 20°C, 0.5 cm³ of oxygen was collected. Assuming no experimental error, explain why this small amount of oxygen was produced. *(2 marks)*

(c) (i) Explain the difference in the volumes of oxygen collected at 20°C and at 30°C. *(2 marks)*

(ii) Explain the difference in the volumes of oxygen collected at 35°C and at 50°C. *(2 marks)*

Total: 9 marks

5 (a) The graph below shows the activity of three enzymes at different temperatures. Each enzyme comes from a different organism.

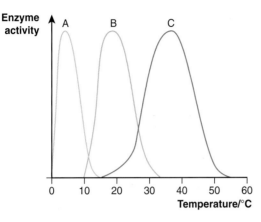

Suggest an explanation for the different activities of the enzymes. *(4 marks)*

(b) A metabolic pathway in which each stage is controlled by enzymes is represented as:

$$A \xrightarrow{e_1} B \xrightarrow{e_2} C \xrightarrow{e_3} D \xrightarrow{e_4} E$$

(i) Use this example to explain what is meant by end-product inhibition. *(3 marks)*

(ii) Explain the induced-fit model of enzyme action. *(3 marks)*

Total: 10 marks

Chapter 4

Digested food must be absorbed into the bloodstream

This chapter covers:
- the structure of plasma membranes
- diffusion
  - simple diffusion
  - facilitated diffusion
- active transport
- osmosis

Food that has been digested must pass into the bloodstream so that it can be circulated around the body to be used. The food molecules must therefore pass into and then out of the epithelial cells lining the small intestine before they pass through capillary walls and into the bloodstream. This means that they must pass through the plasma membranes of the epithelial cells. Different processes are involved in transporting different molecules across the plasma membranes.

Box 4.1 Particles, atoms, ions and molecules

'Particle' is a generic term that covers atoms, ions and molecules.

An atom is an uncharged particle. It is the smallest particle of an element that can take part in a chemical reaction (e.g. an oxygen atom or an iron atom). An ion is a charged particle. It may be just one atom that has become charged (e.g. oxide, O^{2-} or iron(II), Fe^{2+}) or a group of atoms that is charged (e.g. hydrogen carbonate, HCO_3^-). A molecule is an uncharged particle made from several atoms, which may be the same (e.g. an oxygen molecule, O_2) or different (e.g. a copper sulphate molecule, $CuSO_4$).

All particles have a certain amount of **kinetic energy** that causes them to move in a random manner. Sometimes, movement across membranes is the direct result of random movement of particles.

 When you are describing the motion of particles, be careful not to use phrases such as 'the particles *need* to move into …' or 'the cell *needs* the particles to move into it…'. Needing is an emotion or a desire. Cells and particles do not have emotions or desires!

How do the digested food molecules in the gut lumen reach the capillaries in the gut wall?

To reach the capillaries in the gut wall, digested food molecules must pass through the epithelial cells that line the small intestine. This is illustrated in Figure 4.1 (see also Figure 3.3 on page 47).

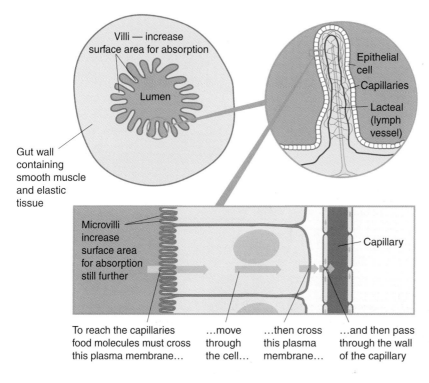

Figure 4.1 How digested food molecules reach the capillaries in the gut wall

You will see that to reach the capillaries, food molecules must cross two plasma membranes:
- the plasma membrane of the epithelial cells next to the lumen, which is folded into microvilli to increase the surface area for absorption
- the plasma membrane of the epithelial cells away from the lumen and closer to the capillaries

To understand how this happens, we must first understand the structure of plasma membranes.

How are plasma membranes put together?

All the current evidence supports the **fluid mosaic model** of membrane structure.

The fluid mosaic model is so named because it suggests that:
- the phospholipid molecules are not fixed absolutely in one position, but move (as do the particles in a fluid), although the movement is mainly lateral (not in all directions as in a fluid); the protein molecules move to a much lesser extent
- when either surface is viewed, there is a 'mosaic' pattern made from the protein molecules and the 'heads' of the phospholipids

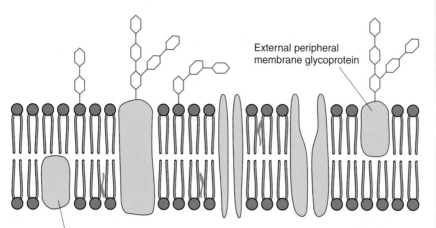

External peripheral membrane glycoprotein

Internal peripheral membrane protein

Figure 4.2 The fluid mosaic model of the structure of a plasma membrane

The different proteins in the plasma membrane have different functions. Many of the **integral (transmembrane) proteins** are involved in moving particles across the membrane. They can be:

- **ion pores** or channels — to allow the passage of ions across the membrane
- **transport proteins** — to carry molecules such as glucose and amino acids through the membrane, sometimes passively and sometimes using energy from ATP

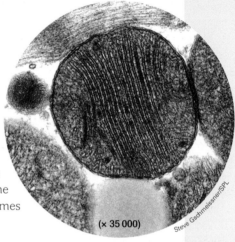

False-colour transmission electron micrograph of a mitochondrion, showing the membrane appearance

(× 35 000)

Steve Gechmeissner/SPL

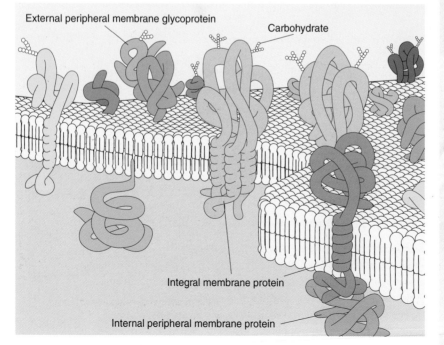

External peripheral membrane glycoprotein

Carbohydrate

Integral membrane protein

Internal peripheral membrane protein

Figure 4.3 Three-dimensional diagram of membrane structure

Be sure that you know the *functions* of the various components of the plasma membrane, as well as their names. You may be asked, for example, to 'identify the component that is responsible for cell recognition', rather than just to 'identify a peripheral protein'.

Integral proteins can also be enzymes, proteins involved in energy transfer and support structures.

Peripheral proteins in the inner phospholipid layer may help to anchor other proteins in the membrane. Those in the outer phospholipid layer often have carbohydrate chains attached, i.e. they are glycoproteins. These are important in cell recognition and as receptors for molecules such as hormones.

Box 4.2 The development of the fluid mosaic model of membrane structure

In the late nineteenth century, experiments showed that cell membranes allowed lipid-soluble substances to enter cells more quickly than water-soluble substances. This led biologists to suggest that the plasma membrane was made entirely from lipid. In the early twentieth century, chemical analysis showed that these lipids were phospholipids and experiments showed that these could be made to form a bilayer. However, these models could not account for the stability of the membrane in a watery medium.

By 1930, further analysis and experiments showed that there was also protein present in the membranes and that water-soluble substances entered faster than should be possible if the membrane were pure lipid. In 1934, Davson and Danielli proposed a model of membrane structure in which a phospholipid bilayer was 'sandwiched' between two layers of protein. Early electron micrographs showed plasma membranes as three-layered structures with a lighter region sandwiched between two darker regions. This seemed to support the model.

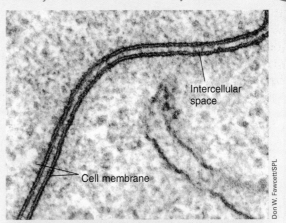

Intercellular space

Cell membrane

However, this model could not account for the fluid nature of the membrane, which had been noticed for many years and was being demonstrated by more and more experiments. Scanning electron micrographs were able to show that protein molecules did not just sit on the surface but, in many cases, were an integral part of the membrane. In 1972, Singer and Nicholson proposed the fluid mosaic model to account for this. This model proposes that the molecules making up the membrane can flow freely in the plane of the membrane — the membrane is a fluid. However, they are fixed in position with regard to the depth of the membrane. The 'mosaic' part of the name comes from images of proteins in the membrane, forming a 'mosaic' pattern.

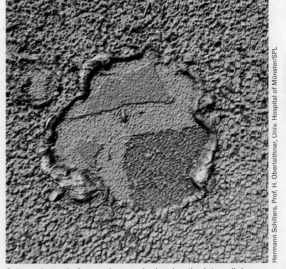

Coloured atomic force micrograph showing the intracellular surface of a plasma membrane

The way in which the changing evidence influenced biologists' models of membrane structure is summarised in the table.

Evidence	Influence on model of membrane structure
Lipid-soluble substances enter cells faster than water-soluble substances	Model of plasma membrane made entirely from lipids
Chemical analysis reveals lipids to be phospholipids and experiments show that phospholipids form bilayers	Model modified to suggest a bilayer of phospholipid
Further analyses show protein is also present. Further experiments show that water enters cells more quickly than is possible for a pure lipid membrane	Davson and Danielli model of a phospholipid bilayer sandwiched between two layers of protein
Electron micrographs show membrane as a three-layered structure	Davson–Danielli model supported
More and more evidence of fluid nature of membrane; scanning electron micrographs reveal that proteins are an integral part of membrane structure	Fluid mosaic model developed – phospholipid bilayer with protein molecules integral to the membrane structure; all molecules can move in plane of membrane

How do particles cross plasma membranes?

Diffusion

Simple diffusion is defined as 'the net movement of particles from an area of high concentration to an area of lower concentration'.

This can only take place in a fluid (a liquid or a gas). In a solid, the particles are fixed in position and cannot move to allow diffusion to take place. Simple diffusion does not necessarily involve a membrane. When someone sprays air freshener in a room, the particles of the freshener diffuse from the point of spraying (the high concentration) to the rest of the room (the lower concentration).

Diffusion *can* occur across a membrane, provided the particles are physically able to cross it. A plasma membrane contains a phospholipid bilayer and the 'tails' of the phospholipids are non-polar, i.e. they are not charged. This makes it difficult for ions (charged particles) to pass through. Also, the phospholipids are tightly packed, which means that large molecules are unlikely to be able to pass through. Table 4.1 shows why some substances can diffuse through the plasma membrane and why some cannot.

◀ 'Net' means 'on balance'. Particles move in both directions, but more move from the high concentration to the low concentration than move in the other direction. *On balance*, particles appear to move only from the high to low concentration.

◀ Diffuse means spread out.

◀ If a substance is lipid-soluble, its particles can mix with those of a lipid. This allows such particles to pass between the molecules of phospholipid in the plasma membrane.

Substance	Type of particle	Is the particle charged?	Size of particle	Is the particle lipid-soluble?	Can the particle diffuse through the membrane?
Water	Molecule	No	Small	No	Yes
Glycerol	Molecule	No	Small	Yes	Yes
Fatty acids	Molecule	Slightly	Medium	Yes	Yes
Carbon dioxide	Molecule	No	Small	No	Yes
Oxygen	Molecule	No	Small	No	Yes
Glucose	Molecule	No	Medium	No	No
Sodium chloride	Ions (Na⁺, Cl⁻)	Yes	Small	No	No
Protein	Molecule	No	Large	No	No

Table 4.1 Types of particle that can or can't diffuse through the plasma membrane

◀ In the small intestine, the only particles to pass through the plasma membranes of the epithelial cells by simple diffusion are fatty acids and glycerol. These molecules are lipid-soluble and so can pass directly through the phospholipid bilayer.

If the particles of a substance are lipid-soluble, even if the particles are quite large, they can diffuse through the plasma membrane. If the particles are not lipid-soluble, diffusion through the plasma membrane is only possible if they are small and non-polar (uncharged). Plasma membranes are, therefore, **partially permeable** — they allow some particles to pass freely, but not others.

Factors that affect the rate of diffusion
Concentration gradient

For *net* diffusion to occur, there must be a concentration difference of substance on either side of a membrane. This difference in concentrations is called a **concentration gradient**. There is a *net* movement from the higher concentration to the lower concentration. This continues until **equilibrium** is reached. At equilibrium, the concentrations on each side of the membrane are equal (the concentration gradient is zero) and there is equal diffusion in both directions (there is no net diffusion).

The greater the difference in the two concentrations, the faster the initial rate of net diffusion will be. The rate of net diffusion slows down as the two concentrations become closer to each other.

◀ When a concentration gradient is present, diffusion occurs spontaneously.

e In an examination, you need not use the word 'net'. The examiner will assume that you are describing the net movement of particles, unless you state otherwise.

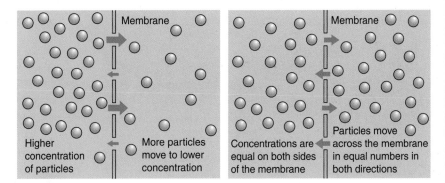

| Higher concentration of particles | More particles move to lower concentration | Concentrations are equal on both sides of the membrane | Particles move across the membrane in equal numbers in both directions |

Figure 4.4 The concentration gradient affects the rate of diffusion

The rate of diffusion of a substance across a membrane (or exchange surface) is proportional to the difference in concentration across the membrane (or exchange surface).

Thickness of surface

The distance that particles have to travel affects how long it takes for them to cross the membrane and therefore affects the rate of diffusion. All plasma membranes are essentially the same thickness, but the layers of cells in different exchange surfaces are not. A thin exchange surface allows faster diffusion than a thick one.

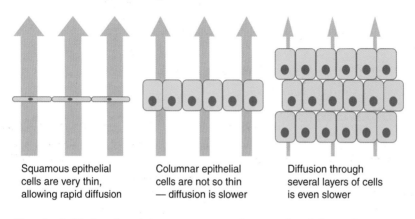

Squamous epithelial cells are very thin, allowing rapid diffusion

Columnar epithelial cells are not so thin — diffusion is slower

Diffusion through several layers of cells is even slower

Figure 4.5 The thickness of an exchange surface affects the rate of diffusion

◀ Diffusion also occurs through *layers* of cells called **exchange surfaces**. For example:
● fatty acids and glycerol diffuse through epithelial cells in the small intestine
● oxygen and carbon dioxide diffuse through epithelial cells in the alveoli
● urea diffuses through epithelial cells in kidney tubules

The rate of diffusion of a substance across an exchange surface is inversely proportional to the thickness of the exchange surface.

Surface area

Another factor that affects the rate of diffusion across a membrane or exchange surface is the surface area. A large surface area allows faster diffusion (more particles will cross per second) than a small surface area. Cells specialised for absorption frequently have a plasma membrane with microvilli to increase the surface area.

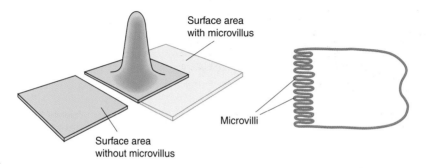

Surface area with microvillus

Microvilli

Surface area without microvillus

Figure 4.6 Microvilli increase the surface area of a plasma membrane

The rate of diffusion of a substance across a membrane (or exchange surface) is proportional to the surface area of the membrane (or exchange surface).

Fick's law

The effect of all these factors on diffusion is summarised in **Fick's law**:

$$\text{rate of diffusion} \propto \frac{\text{surface area} \times (C_2 - C_1)}{\text{thickness of exchange surface}}$$

where $(C_2 - C_1)$ = difference in concentration

In summary, simple diffusion describes the movement of particles down a concentration gradient as a direct result of random motion. It is a passive process, requiring no extra energy in the form of ATP from respiration. The only energy involved is the kinetic energy of the molecules themselves.

Facilitated diffusion

Particles that cannot cross membranes by simple diffusion — for example, sodium ions, glucose and amino acid molecules — often enter or leave cells by **facilitated diffusion**. This process still relies on the presence of a concentration gradient across the membrane, down which the particles diffuse. However, the particles do not pass freely between the molecules of phospholipid. Movement is facilitated (made possible) by protein molecules that span the plasma membrane.

Channel proteins are made from several subunits and have a hole or pore running through the centre. They are responsible for the transport of ions into and out of cells and form **ion pores** or **ion channels**. Some are specific and only allow certain ions to pass through (e.g. sodium-ion channels, potassium-ion channels, calcium-ion channels); others are non-specific and allow several different ions to pass through. Some are open all the time whereas others have gates (**gated channels**) that are opened and closed by appropriate stimuli, such as pressure, or a change in voltage.

◀ A channel protein can be thought of as a tube that spans the membrane.

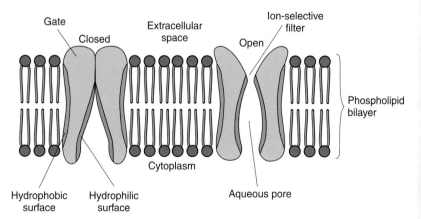

Figure 4.7 Different states of a gated ion channel

Carrier proteins transport medium-sized molecules into and out of cells. To facilitate the movement of the diffusing molecule, the carrier protein molecule usually has to undergo a conformational change (change in shape).

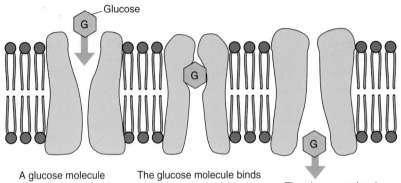

Glucose

A glucose molecule diffuses into the carrier protein molecule

The glucose molecule binds with the carrier protein molecule, causing the protein molecule to change shape

The glucose molecule diffuses out of the carrier protein molecule

Figure 4.8 Facilitated transport of glucose across a plasma membrane

As with simple diffusion, there is net movement of particles across the membrane only when a concentration gradient is present.

Some carrier proteins function in **co-transport systems**. One of these is responsible for the uptake of sodium ions and glucose molecules in the small intestine (and also in the nephrons in the kidney). A similar system operates for the co-transport of sodium ions and amino acids.

◀ Glucose molecules *enter* the epithelial cells lining the small intestine through sodium–glucose co-transport proteins.

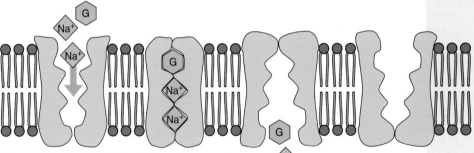

When sodium ions bind to the transport protein molecule a change in shape allows the glucose to bind

The binding of glucose closes the pore to the outside and opens the pore to the inside, releasing the sodium ions and glucose into the cell

The release of the sodium ions and glucose reopens the pore to the outside

Figure 4.9 Co-transport of sodium ions and glucose molecules

Box 4.3 Membrane proteins and the rate of diffusion

The presence of channel proteins and carrier proteins greatly affects the rate of diffusion of particles into and out of cells. For example, the rate of diffusion of chloride ions across the membrane of a red blood cell is increased by a factor of 10^7 and the permeability of most cells to glucose is increased by a factor of 50 000.

However, the presence of channel and carrier proteins does impose a finite limit on the rate of facilitated diffusion. Each protein can only transport a certain number of particles per second. Increasing the concentration gradient increases the rate of facilitated diffusion up to the point at which all the carrier proteins or ion channels are transporting at their maximum rate. Facilitated diffusion cannot then take place any faster. Any further increase in the concentration gradient has no effect.

With simple diffusion, this is not the case. The initial rate is slower, but as the concentration gradient is increased, the rate of simple diffusion just keeps increasing.

The relationship between concentration gradient and the rates of uptake of a solute by simple and facilitated diffusion are shown in Figure 4.10

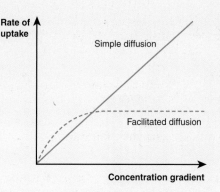

Figure 4.10 The change in rates of simple diffusion and facilitated diffusion as the concentration gradient increases

Fick's law has to be modified slightly for facilitated diffusion. Surface area in itself does not influence the rate of facilitated diffusion, but the number of transport or channel protein molecules does. The law must be rewritten as:

$$\text{rate of diffusion} \propto \frac{\text{number of transport protein molecules} \times (C_2 - C_1)}{\text{thickness of exchange surface}}$$

Since the exchange surface is always a plasma membrane, the rate is really only influenced by the number of transport proteins and the difference in concentration.

Active transport

Sometimes, substances must be moved *against* a concentration gradient — from a low concentration to a higher one. This cannot happen by diffusion, since it would tend to concentrate particles rather than spread them out. It can only happen if energy is put in to drive the process. In living organisms, this energy is released from the ATP produced in respiration. The proteins used to actively transport substances across plasma membranes are called **pumps**.

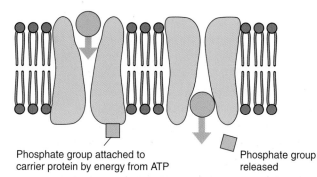

Phosphate group attached to carrier protein by energy from ATP

Phosphate group released

Phosphate group activates carrier protein to accept particle to be transported

Energy from release of phosphate group used to change shape of carrier protein — transporting particle across membrane

Figure 4.11 Role of ATP in active transport

There is a protein pump that transports sodium ions out of a cell at the same time as it transports potassium ions into a cell. This protein is known as the **sodium–potassium pump**. The configuration of the protein is such that three sodium ions are transported out of the cell while only two potassium ions are brought in.

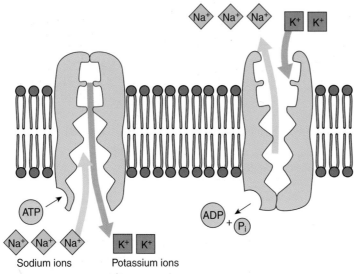

Figure 4.12 The sodium-potassium pump

ATP binds to the protein to allow Na⁺ to enter the pump and K⁺ to leave

Hydrolysis of ATP with the release of energy allows Na⁺ to leave the pump and K⁺ to enter

Box 4.4 How does glucose reach the blood plasma from the lumen of the small intestine?

The glucose must cross three barriers:
- the plasma membrane of the epithelial cells next to the lumen of the small intestine
- the plasma membrane of the epithelial cells away from the lumen and closer to the capillaries
- the wall of the capillaries

A sodium–potassium pump actively transports sodium ions out of the epithelial cell. This reduces the concentration of sodium ions in the cell, which, in turn:
- creates a concentration gradient between the lumen of the small intestine and the cell
- allows more sodium ions to enter through the co-transport protein

However, for a sodium ion to enter a cell, a molecule of glucose must also enter. These glucose molecules then pass by facilitated diffusion out of the cell, down a concentration gradient, into the intercellular space between the epithelial cells and the capillary. They then pass by *simple* diffusion through the capillary wall into the blood plasma. The concentration gradient between the intercellular space and blood is maintained by the flow of blood. Without the active transport of sodium ions out of the epithelial cell, the uptake of glucose would be considerably reduced.

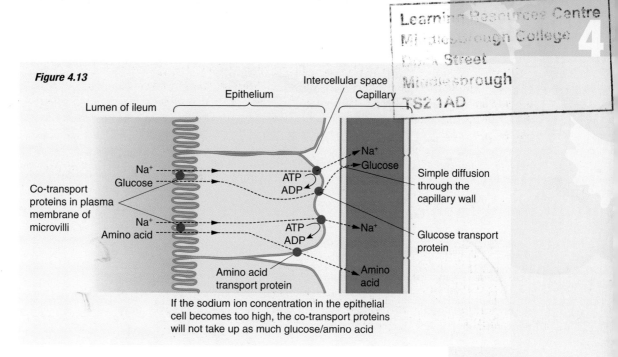

Figure 4.13

If the sodium ion concentration in the epithelial cell becomes too high, the co-transport proteins will not take up as much glucose/amino acid

Osmosis

Osmosis is the process by which water moves across a partially permeable membrane. It is, effectively, the diffusion of water. However, we do not refer to the concentration of water molecules, but to **water potential**. Water potential is a measure of the free energy of the water molecules in a system. Therefore, water moves, by osmosis, from a system with a high water potential to a system with a low water potential. The symbol for water potential is the Greek letter Ψ (psi). Water potential is measured in units of pressure — pascals (Pa), kilopascals (kPa), or megapascals (MPa). Pure, liquid water has a higher water potential than any other system. It is defined as zero:

$$\Psi(\text{pure water}) = 0 \text{ Pa}$$

All other systems (cells, solutions and suspensions) have a water potential that is lower than that of water. Therefore, their water potential values are negative.

Osmosis is the movement of water from a system with a high (less negative) water potential to one with a lower (more negative) water potential, across a partially permeable membrane.

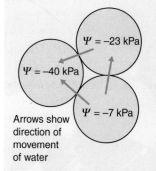

Arrows show direction of movement of water

Figure 4.14 Water movement between cells with different water potentials

Why water potential values are negative

The water potential of a system is due to the number of water molecules in that system. In pure water, there are only water molecules. When a solute is added, some of the water molecules form 'hydration shells' around the solute molecules. This reduces the number of (free) water molecules in the system and so the water potential is reduced. Since pure water is assigned a water potential of zero, the solution must have a negative water potential. A more concentrated solution will take more free water molecules out of the system and lower the water potential still further, making it more negative.

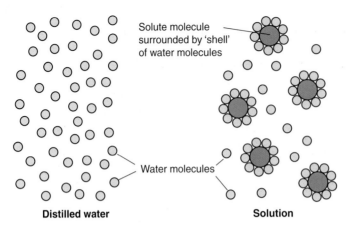

Figure 4.15 Adding a solute reduces the water potential of a system

Cells contain many molecules in an aqueous system. The sugars, ions and proteins present in a cell give it a negative water potential. Different concentrations of these substances result in cells with different water potentials. Water will move between them until the water potentials are all the same.

Box 4.5 Oral rehydration therapy

People with diarrhoea can become dehydrated easily. This used to kill millions of people in developing countries each year.

Dehydration can be treated by putting the patient on a 'drip'. A solution containing ions and sugars is dripped into the blood plasma. This both adds fluid and makes the blood plasma slightly more concentrated and causes the absorption of water from the small intestine by osmosis. This does not cure the diarrhoea, but prevents the dehydration that is the killer.

However, in developing countries there are insufficient trained personnel to administer this treatment on the scale that is needed. In the 1970s, after finding out that glucose and ions could still be reabsorbed by the small intestine, doctors working in Pakistan and Bangladesh experimented with administering the solutions to patients orally (by mouth).

They found that a simple solution of one teaspoon of salt and four teaspoons of sugar in 1 litre (1 dm³) of water had the same effect as a drip. The absorption of the salt and sugar caused water to be absorbed (by osmosis). This prevented dehydration.

More recently it has been shown that including amino acids in the solution further increases rehydration. This happens because amino acids are co-transported from the small intestine with glucose. The glucose can be given in the form of starch which, over a period of time, is digested to glucose. This is a way of getting more glucose to the patient without giving it as a concentrated solution — which could make the diarrhoea worse by causing water to be lost from the blood by osmosis.

There are commercially prepared oral rehydration therapy packages that contain precisely controlled amounts of sodium and potassium salts, glucose (as starch) and amino acids. However, including half a mashed banana with the salt and sugar solution achieves almost the same end. Using rice powder in the solution is another way of providing starch and amino acids.

In developing countries, diarrhoea is common, often occurring as a result of cholera (see Chapter 7).

Were the doctors justified in experimenting on humans? They may have saved millions of lives, but what about the people who were the 'guinea pigs'?

Transport processes compared

The transport processes that occur in living organisms are compared in Table 4.2.

Process	Influence of concentration	Energy requirement	Type of particles moved across the plasma membrane	Transport proteins
Simple diffusion	Occurs down a concentration gradient — from high to low	Kinetic energy of the particles — passive process	Lipid-soluble, small, non-polar particles	None
Facilitated diffusion	Occurs down a concentration gradient — from high to low	Kinetic energy of the particles — passive process	Ions and medium-sized particles that are insoluble in lipid (e.g. glucose)	Channel proteins or carrier proteins
Active transport	Occurs against a concentration gradient — from low to high	ATP from respiration in addition to kinetic energy of the particles — active process	Ions and medium-sized particles that are insoluble in lipid (e.g. glucose)	Pumps
Osmosis	Occurs down a water potential gradient — from high to low (less negative to more negative)	Kinetic energy of the particles — passive process	Water	None

Table 4.2

The kinetic energy of the particles is responsible for their movement. Therefore, the rates of all the processes will potentially increase with rising temperature — as the particles gain more kinetic energy. However, increasing the temperature too much will denature transport proteins, which will reduce the rates of facilitated diffusion and active transport.

> ℓ In an examination, do not write that *energy* is not needed for diffusion to take place. *ATP* is not needed for diffusion to take place, but diffusion could not occur without the kinetic energy of the particles of the diffusing substance. The same is true of facilitated diffusion and osmosis.

Summary

Plasma membrane structure

- The fluid mosaic model of membrane structure suggests a fluid phospholipid bilayer with interspersed protein molecules, forming a mosaic pattern.
- The two main types of protein are integral (transmembrane) proteins that span the membrane and peripheral proteins that occur only on one surface of the phospholipid bilayer.

- Integral proteins are often involved in transport. There are ion channels/pores and transport proteins (both involved in facilitated diffusion), and pumps that are used in active transport.

Transport processes

- Substances are transported into and out of cells by simple diffusion, facilitated diffusion, active transport and osmosis,. Transport through a *layer* of cells may involve more than one of these processes.
- Simple diffusion is the movement of particles down a concentration gradient (from a high concentration to a lower concentration). Transport proteins and ATP are not needed. Simple diffusion can occur across a membrane but often occurs in systems where no membrane is present.
- Facilitated diffusion is the movement of particles across a membrane through a channel protein or carrier protein. Movement is down a concentration gradient and ATP is not needed.
- Active transport is the movement of particles across a membrane through pumps (ATP-dependent transport proteins). Movement is against a concentration gradient and ATP is required.
- Osmosis is the movement of water across a partially permeable membrane from a high (less negative) water potential to a lower (more negative) water potential. Movement is, therefore, down a water potential gradient. Transport proteins and ATP are not needed.
  - Animal cells placed in a solution of higher water potential take in water by osmosis and burst because the plasma membranes cannot withstand the internal pressure.
  - Plant cells placed in a solution with a higher water potential take in water by osmosis but do not burst because the cell wall is able to withstand the internal pressure. The cells become turgid.

Factors affecting diffusion rate

- The rate of diffusion across an exchange surface (which may be a plasma membrane or a specialised exchange surface one or more cell layers thick) is described by Fick's law:

$$\text{rate of diffusion} \propto \frac{\text{surface area} \times (C_2 - C_1)}{\text{thickness of exchange surface}}$$

where $(C_2 - C_1)$ is the difference in concentration between the solutions on each side of the exchange surface.
- The rate of diffusion is increased by:
  - increasing the surface area
  - increasing the concentration gradient
  - decreasing the thickness of the exchange surface
  - increasing the temperature
- The absorption of glucose from the small intestine involves:
  - a glucose-sodium co-transport protein that transports both particles into the epithelial cells from the lumen of the gut

- active transport of sodium ions out of the epithelial cell to maintain a concentration gradient of sodium ions between the lumen of the small intestine and the epithelial cell
- facilitated diffusion of glucose out of the epithelial cell into the intracellular space
- simple diffusion of glucose from the intercellular space into the blood plasma

Questions

Multiple-choice

1 The fluid mosaic model proposes that plasma membranes consist of:
   A a protein bilayer with phospholipids interspersed between the protein molecules
   B two layers of lipid with protein between them
   C two layers of protein with lipid between them
   D a phospholipid bilayer with protein molecules interspersed between the phospholipid molecules

2 Proteins in the plasma membrane may be involved in:
   A transport of ions in and out of cells through ion pores
   B cell recognition
   C the sodium–potassium ion pump
   D all of the above

3 Substances that prevent respiration from occurring also prevent active transport from taking place because:
   A there will be no oxygen available
   B there will be no ATP available
   C there will be too much carbon dioxide
   D all of the above

4 Diffusion always involves the movement of particles:
   A out of a cell
   B into a cell
   C down a concentration gradient
   D against a concentration gradient

5 Amino acid molecules are removed from an epithelial cell in the small intestine by:
   A facilitated diffusion
   B active transport
   C a combination of facilitated diffusion and active transport
   D some other process

6 Active transport always involves:
   A movement against a concentration gradient
   B a transport protein acting as a pump
   C the release of energy from ATP
   D all of the above

7 Ions cannot pass across a plasma membrane by simple diffusion because the particles are:
  A too large
  B non-polar
  C hydrophobic
  D charged

8 When placed in a solution that has a more negative water potential, animal cells:
  A lose water by osmosis and shrink
  B take in solutes from the solution and shrink
  C take in water by osmosis, swell and burst
  D lose solutes to the solution and swell

9 Facilitated diffusion always involves:
  A movement of ions
  B movement against a concentration gradient
  C ATP
  D transport proteins

10 As the concentration gradient is increased:
  A the rate of both simple diffusion and facilitated diffusion will increase continually
  B the rate of simple diffusion will rise and then level off but the level of facilitated diffusion will increase continually
  C the rate of simple diffusion will increase continually but the rate of facilitated diffusion will rise and then level off
  D the rate of both simple diffusion and facilitated diffusion will rise and then level off

Examination-style

1 The diagram below shows the structure of part of a plasma membrane as seen in section.

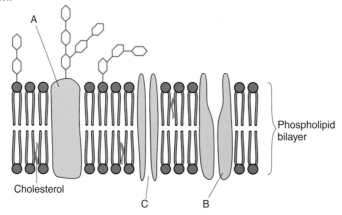

(a) Name the structures labelled A and B. *(2 marks)*
(b) Describe the function of the structure labelled C. *(2 marks)*
(c) Explain why phospholipid molecules in the membrane are orientated in the way shown in the diagram. *(3 marks)*

Total: 7 marks

2 The graph below shows the change in the rate of uptake of two substances, A and B, by a cell as the external concentration of each changes. Initially, the internal concentration of both substances is the same. One of the substances is absorbed by simple diffusion, the other by facilitated diffusion.

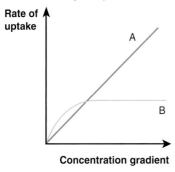

(a) Describe how the two rates are:
  (i) similar   (*1 mark*)
  (ii) different   (*1 mark*)
(b) Which substance, A or B, is absorbed by facilitated diffusion? Explain how this can be deduced from the evidence in the graph.   (*4 marks*)
(c) Explain how a co-transport protein functions.   (*4 marks*)

Total: 10 marks

3 An investigation was carried out into the permeability of plasma membranes of beetroot cells. These cells contain a purple-red pigment that, ordinarily, can only pass through the plasma membranes to a very limited extent. The procedure was as follows:

- Remove some cylinders of beetroot using a cork borer.
- Cut ten discs of beetroot, 5 mm thick.
- Wash the discs until no more colour escapes.
- Place the discs in a test tube containing 10 cm^3 distilled water, which has been allowed to equilibrate in a water bath at 20°C for 10 min.
- Replace the test tube in the water bath for 20 min.
- Swirl the tube and pour some of the liquid into a cuvette.
- Measure the intensity of the colour of the water by placing the cuvette in a colorimeter, set to read absorbance.
- Repeat twice.
- Repeat the whole procedure at temperatures other than 20°C.

The results obtained are represented in the graph below.

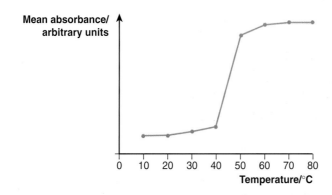

(a) (i) Suggest why there was some coloration of the water even at low temperatures.   (*2 marks*)
  (ii) Explain why it was necessary to wash the beetroot discs.   (*2 marks*)

(b)(i) Describe the change in absorbance as the temperature increases. *(3 marks)*

(ii) Use the fluid mosaic model of membrane structure to explain the changes described in (b)(i). *(3 marks)*

Total: 10 marks

4 The diagram below represents the uptake of glucose, sodium ions and water by a cell in a kidney tubule.

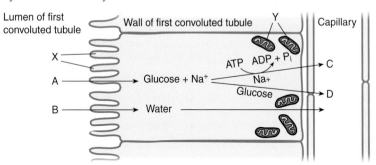

(a) Use Fick's law to explain how the structures labelled X help to maintain a high rate of absorption from the lumen of the kidney tubule. *(3 marks)*

(b) Explain how the presence of the organelles labelled Y is essential to the absorption of glucose. *(4 marks)*

(c) Copy and complete the table below by placing a tick or a cross in each box to indicate whether the feature does or does not apply to each process. *(3 marks)*

Process	Feature of process		
	Substance moves against concentration gradient	Process is passive	Process requires carrier proteins
A			
B			
C			
D			

Total: 10 marks

5 Plasma membranes consist of a phospholipid bilayer in which protein molecules are interspersed.

(a) The table below shows the percentage masses of protein, lipid and carbohydrate in four different plasma membranes.

Membrane	Percentage mass		
	Protein	Lipid	Carbohydrate
A	18	79	3
B	51	49	0
C	52	44	4
D	76	24	0

(i) Calculate the mean ratio of protein to lipid for the
four plasma membranes. *(2 marks)*

(ii) Describe two functions of protein molecules in plasma
membranes. *(2 marks)*

(iii) Describe one function of carbohydrates in plasma
membranes. *(1 mark)*

(iv) Suggest why plasma membrane D has a much higher
protein content than plasma membrane A. *(2 marks)*

(b) When phospholipid bilayers are heated, the phospholipid 'tails' become
more mobile. At a critical transition temperature, they absorb a great deal
of heat and become so mobile that they behave like a liquid. The graph
below shows the effect of temperature on the heat absorption of a pure
phospholipid bilayer and one to which 20% cholesterol has been added.

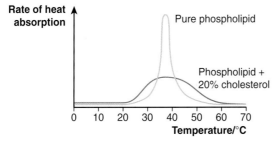

(i) How would the phospholipid tails behaving like a
liquid affect the permeability of a plasma membrane?
Explain your answer. *(3 marks)*

(ii) Over what range of temperatures does the pure
phospholipid bilayer undergo the transition where
the phospholipid tails become liquid? *(1 mark)*

(iii) How is this different for the bilayer with 20% cholesterol
added? *(2 marks)*

(iv) Suggest a function for cholesterol molecules in
plasma membranes. Explain your answer. *(2 marks)*

Total: 15 marks

6 Scientists have produced a mathematical model of the uptake of glucose in
the small intestine. By varying factors in the model, they can simulate the
effect on glucose uptake of different levels of exercise following a meal.

(a) (i) With respect to investigating the effects of exercise on
glucose uptake, suggest two benefits of using a
mathematical model of glucose absorption, rather than
carrying out experiments. *(2 marks)*

(ii) Suggest two disadvantages of using the mathematical
model. *(2 marks)*

(b) The model predicts the following results for the effect of exercise on
glucose concentration in the portal vein after different levels of exercise.
The portal vein carries blood away from the small intestine. The exercise
levels were simulated by varying the estimated blood flow to the small
intestine in the mathematical model.

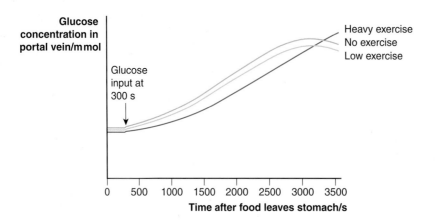

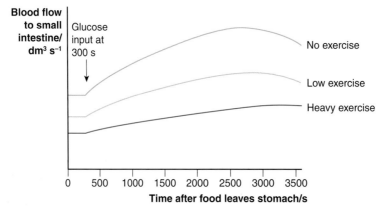

(i) Explain how the information given concerning the
   level of exercise makes the data difficult to interpret. *(3 marks)*

(ii) Suggest how varying the blood flow to the small
   intestine in the way shown could simulate different
   levels of exercise. *(3 marks)*

(iii) Describe the effect of heavy exercise on glucose uptake. *(2 marks)*

(c) Most of the uptake of glucose from the small intestine depends
   on active transport. Why might a change in blood flow to the
   small intestine affect the rate of glucose uptake? *(3 marks)*

Total: 15 marks

Chapter 5

The circulatory system transports absorbed food around the body

This chapter covers:
- a brief introduction to the mammalian circulatory system
- the structure of the heart
- how the heart pumps blood through the circulatory system
- how the heartbeat is controlled

The circulatory system is one of the main organ systems found in a mammal. It links other organ systems by transporting substances between them. For example, the products of digestion are absorbed in the small intestine and then distributed to other organs. Surplus glucose in the blood plasma is taken to the liver and stored as glycogen. At the centre of this system is the heart, which is a muscular pump. Contractions of the ventricles of the heart generate the force to pump blood through the blood vessels that comprise the circulatory system. Although the heartbeat is myogenic, it is influenced by both nerves and hormones.

The heartbeat is said to be **myogenic** because it originates in the heart muscle itself. The prefix 'myo' always denotes muscle.

What is the circulatory system?

Mammals have a **double circulation**. During a complete circulation of the body, the blood passes through the heart twice. It is pumped to the lungs to be oxygenated (the **pulmonary circulation**) and then returns to the heart to be pumped to other parts of the body that use the oxygen (the **systemic circulation**).

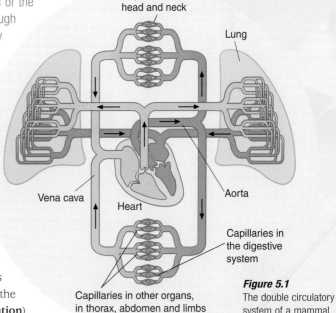

Capillaries in head and neck

Lung

Vena cava

Heart

Aorta

Capillaries in the digestive system

Capillaries in other organs, in thorax, abdomen and limbs

Figure 5.1
The double circulatory system of a mammal

Two major advantages of a double circulatory system over a single circulatory system are that:

- blood passing to the tissues is always oxygenated (saturated with oxygen) because the pulmonary and systemic circulations are separate
- blood is delivered to the tissues at high pressure (producing a more efficient circulation) because it is pumped twice by the heart

Our never-tiring pump

At the centre of the circulatory system is a muscular pump — the heart. This organ supplies the force to pump blood to all parts of the body. It is made largely from cardiac muscle but also contains **Purkyne tissue** (modified cardiac muscle fibres that conduct impulses), blood vessels, blood and connective tissue.

In an average lifetime of 75 years, the human heart beats approximately 2 750 000 000 times, and never (hopefully!) take a rest.

Box 5.1 The heart's own blood supply

The **cardiac muscle** in the heart wall respires continuously to release the energy needed for contraction. To supply the oxygen and glucose needed, the cardiac muscle has its own blood supply — the coronary circulation. Two **coronary arteries** branch off the aorta just as it leaves the left ventricle. These carry blood into arterioles and the millions of capillaries that supply the cardiac muscle cells. The coronary arteries are narrower than many other arteries and can become blocked more easily. A build-up of atheroma (a mixture of fatty substances including cholesterol) in the coronary arteries narrows them still further. A dislodged blood clot can quite easily block a narrowed coronary artery, causing a coronary thrombosis. This is covered in more detail in Chapter 7.

Coronary arteries

Blockage in artery

The heart is divided into four chambers:

- right and left **atria**, to receive blood returning from the systemic and pulmonary circulations, respectively
- right and left **ventricles**, to force blood through the pulmonary and systemic circulations, respectively

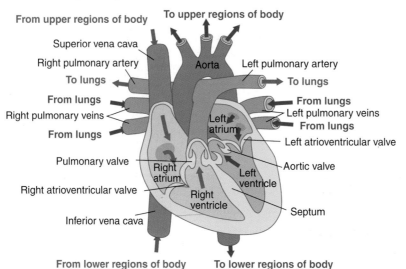

From upper regions of body   To upper regions of body

Superior vena cava
Right pulmonary artery   Aorta   Left pulmonary artery
To lungs   To lungs
From lungs   From lungs
Right pulmonary veins   Left pulmonary veins
From lungs   From lungs
   Left atrium
   Left atrioventricular valve
Pulmonary valve   Aortic valve
Right atrium   Left ventricle
Right atrioventricular valve
Right ventricle   Septum
Inferior vena cava

From lower regions of body   To lower regions of body

Figure 5.2 Structure of the mammalian heart

The left atrioventricular valve is sometimes called the mitral valve, or the bicuspid valve. The right atrioventricular valve is sometimes called the tricuspid valve.

Box 5.2 Valves in the circulatory system

Four valves control the flow of blood in the mammalian heart; one between each atrium and ventricle, and one at the base of each artery leading from the ventricles (Figure 5.2). All the valves are one-way valves and work on essentially the same principle. Blood is a fluid; it flows from an area of high pressure to an area of low pressure. The valves in the heart are designed to open when high pressure is forcing the blood in the 'correct' direction. If high pressure forces blood in the 'wrong' direction, the valves are forced shut.

Blood vessels

There are three basic types of blood vessels:
- **Arteries** carry blood under high pressure away from the heart to the organs.
- **Veins** carry blood under low pressure away from the organs towards the heart.
- **Capillaries** carry blood close to every cell within an organ.

How does the heart beat?

The cardiac cycle

The main events of the cardiac cycle

The four chambers of the heart are continually contracting and relaxing in a definite, repeating sequence called the **cardiac cycle**. When a chamber is contracting, we say it is in **systole**; when it is relaxing, we say it is in **diastole**. So, ventricular systole refers to contraction of the ventricles, whereas atrial diastole refers to relaxation of the atria. The two sides of the heart work together. As the left atrium contracts, so does the right atrium. As the right ventricle relaxes, so does the left ventricle.

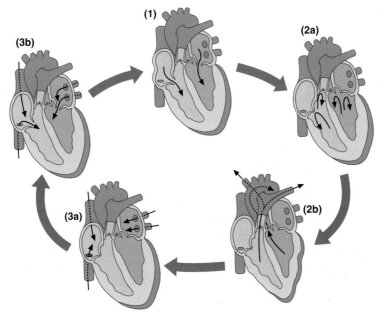

Figure 5.3 The cardiac cycle

Referring to Figure 5.3:

- **Stage 1: Atrial systole/ventricular diastole** The walls of the atria contract. This raises the pressure of the blood in the atria above that in the ventricles and forces open the atrioventricular valves. Blood passes through these valves into the ventricles.

- **Stage 2a: Ventricular systole/atrial diastole** The ventricles contract. This quickly raises the pressure of the blood in the ventricles above that in the atria, and so closes the atrioventricular valves. However, the pressure is still below that in the main arteries, which means that the aortic and pulmonary valves remain closed. Blood does not leave the ventricles.

- **Stage 2b: Ventricular systole/atrial diastole** The ventricles continue to contract. When the pressure of the blood exceeds that in the main arteries, the pulmonary and aortic valves are forced open. Blood is ejected into the pulmonary artery (carrying blood to the lungs) and aorta (carrying blood into arteries that serve all other parts of the body, including the heart itself).

- **Stage 3a: Ventricular diastole/atrial diastole** The ventricles begin to relax and the pressure of the blood quickly falls below that in the main arteries. The higher pressure in these arteries closes the aortic and pulmonary valves.

- **Stage 3b: Ventricular diastole/atrial diastole** As the ventricles continue to relax, the pressure in the ventricles falls below that in the atria. The higher pressure in the atria forces the atrioventricular valves open. Even though the atria are not contracting, blood flows through the open valves by passive ventricular filling.

Pressure changes during the cardiac cycle

The events of the cardiac cycle create pressure changes that are responsible for moving blood through the heart and into the pulmonary and systemic circulations. Figure 5.4 shows the pressure changes in the left atrium, left ventricle and aorta during one cardiac cycle.

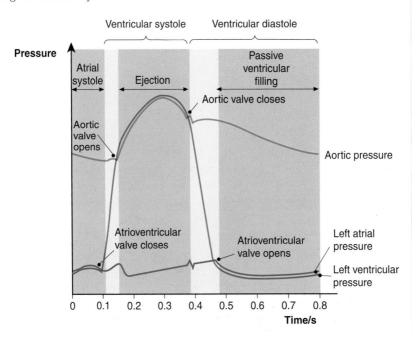

Figure 5.4 The pressure changes during the cardiac cycle

Notice that a valve opens and closes at times in the cycle when the balance of pressures on opposite sides of the valve changes. For example, at 0.1 s, the pressure in the ventricle becomes greater than that in the atrium and so the atrioventricular valve is forced shut. At 0.15 s, the pressure in the ventricle exceeds that in the aorta and so the aortic valve is forced open.

A graph for the changes in the right atrium, right ventricle and pulmonary artery shows all the same features, but the pressures in the right ventricle and pulmonary artery are lower.

> ℮ You may be asked to explain the changes that occur at various points in the cardiac cycle from a graph such as that shown in Figure 5.4. Remember that:
> - atrioventricular valves open as soon as the pressure in the atria becomes greater than that in the ventricles; they close as soon as the pressure in the ventricles becomes greater than that in the atria
> - the aortic and pulmonary valves open as soon as the pressure in the ventricles becomes greater than that in the two arteries; they close as soon as the pressure in the two arteries becomes greater than that in the ventricles.
> - blood flows from a high-pressure region to a low-pressure region.

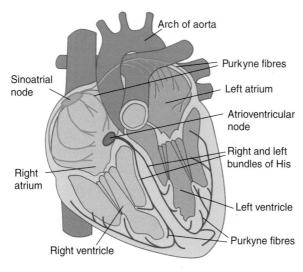

Figure 5.5 The electrical conducting system of the heart

Controlling the cardiac cycle

The events of the cardiac cycle must take place in the correct sequence, with the correct timing. A group of cells in the right atrium form the **sinoatrial node** (SA node), which acts as a natural pacemaker. The SA node initiates electrical impulses that spread through the heart, causing the cardiac muscle to contract. The impulses are carried not by nervous tissue but by specialised, conducting, cardiac muscle fibres called Purkyne tissue. This tissue conducts the impulse throughout the atria, stimulating contraction, to the **atrioventricular node** (AV node). This is the only location where the impulse can pass from atrium to ventricle; all other tissue dividing the two is non-conducting. The impulse is conducted slowly through the AV node. This allows time for the atria to complete their contractions. The impulse then passes along two bundles of Purkyne tissues in the ventricles (the bundles of His) and the cardiac muscle in the ventricles is stimulated to contract. These events are summarised in Table 5.1.

Table 5.1

Event in cardiac electrical conducting system	Event in cardiac cycle	Stage of cardiac cycle
The SA node generates an impulse; the impulse spreads along Purkyne fibres to all parts of the atria	Cardiac muscle in atria contracts, cardiac muscle in ventricles is relaxed — blood is forced through AV valves from atria to ventricles	Atrial systole/ventricular diastole
The impulse is held up at the AV node	Cardiac muscle in atria contracts, cardiac muscle in ventricles is relaxed — blood continues to be forced through AV valves	Atrial systole/ventricular diastole
The impulse is conducted along the bundles of His through the ventricle walls	Cardiac muscle in atria is relaxed, cardiac muscle in ventricles contracts; AV valves closed; then aortic/pulmonary valves opened — blood ejected into main arteries	Atrial diastole/ventricular systole
No impulse	Cardiac muscle in atria and ventricles is relaxed — passive ventricular filling	Atrial and ventricular diastole

Box 5.3 The development of the artificial pacemaker

HOW SCIENCE WORKS

Artificial pacemakers are sometimes implanted into a patient if the natural pacemaker, for some reason, does not generate a sufficiently high heart rate.

- The first attempt to artificially regulate a human heart rate took place in 1889 when J. A. McWilliam reported an experiment in which he had applied electrical impulses at a rate of 60–70 per minute to create a heart rate of 60–70 beats per minute.
- In 1928, Dr M. C. Lidwell used a device plugged into a lighting point to revive a stillborn baby.
- In 1950, a device that plugged into a wall socket was developed by an electrical engineer in Canada. This was painful for the patient and there was a very real risk of electrocution.
- The development of the transistor in 1956 was highly significant because it allowed considerable miniaturisation and, since batteries could now be used, freed the device from an external power source.
- The first implanted pacemaker was developed in Sweden in 1958. It was implanted into the thorax, with electrodes attached to the heart muscle. However, it functioned for only three hours. It was quickly removed; its replacement lasted for two days. The patient, however, survived these tests and lived until 2001, having received over 20 different pacemakers during his lifetime!
- The reliability of these early pacemakers was limited by the power supply. In 1970, this problem was resolved by the development of the lithium-iodide cell.

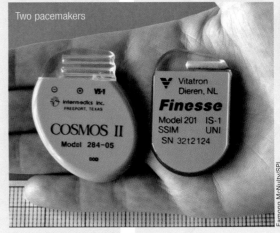

Two pacemakers

Eamonn McNulty/SPL

Cardiac output

Cardiac output is a measure of how hard the heart is working. It is the output from *each* ventricle per minute.

Each time the ventricles contract, they eject blood into the main arteries. The amount of blood ejected from the ventricle is called the **stroke volume** and, at any one time, it is the same for both ventricles. However, this is only one aspect of what determines cardiac output. The other is the **heart rate** — the number of beats per minute. Cardiac output (volume pumped per minute) is the product of the stroke volume (volume per beat) and the heart rate (number of beats per minute).

$$\text{cardiac output} = \text{stroke volume} \times \text{heart rate}$$

> *e* You may be given data about the volume of blood flowing per minute to various organs and asked to calculate the cardiac output. If the data include the volume flowing to the lungs, you need do no more work. The volume flowing to the lungs per minute represents the cardiac output of the right ventricle, which is identical to that of the left ventricle.

An increase in stroke volume or heart rate (or both) increases cardiac output. During exercise, the cardiac output increases to deliver more blood, carrying oxygen and glucose, to the skeletal muscles. During sleep, cardiac output decreases from the normal resting level because the metabolic activity of the body is low and less oxygen is needed by almost all organs.

◀ A cardiac output of 6.5 dm³ means that the right ventricle is pumping 6.5 dm³ into the pulmonary circulation per minute and the left ventricle is pumping 6.5 dm³ into the systemic circulation per minute.

Summary

Structure of the mammalian circulatory system

- Mammals have a double circulatory system. The pulmonary circulation is through the lungs; the systemic circulation is through all other parts of the body.
- The heart is made from cardiac muscle and supplies the force to pump blood through the two circulations.
- The heart comprises four chambers:
  - left atrium, which receives oxygenated blood from the pulmonary circulation via the pulmonary veins
  - left ventricle, which pumps oxygenated blood into the systemic circulation via the aorta
  - right atrium, which receives deoxygenated blood from the systemic circulation via the vena cava
  - right ventricle, which pumps deoxygenated blood into the pulmonary circulation via the pulmonary arteries
- Arteries always carry blood away from the heart.
- Veins carry blood towards the heart.
- Capillaries carry blood to cells within organs.

How the circulatory system of a mammal functions

- In the cardiac cycle:
  - atrial systole forces blood through the atrioventricular valves into the ventricles
  - ventricular systole raises the pressure of blood in the ventricles, which first closes the atrioventricular valves and then opens the pulmonary and aortic valves as blood is ejected into the main arteries
  - atrial and ventricular diastole allow passive ventricular filling
- The heart is myogenic; the SA node generates impulses that are:
  - conducted rapidly along fibres of Purkyne tissue through the atria, stimulating their contraction
  - held up at the AV node, which conducts impulses only slowly
  - conducted rapidly through the ventricles along the bundles of His (also Purkyne tissue), stimulating their contraction
- Cardiac output is the amount of blood pumped per ventricle per minute:
  cardiac output = stroke volume × heart rate

Questions

Multiple-choice

1 The double circulatory system of a mammal comprises:
   A arterial and venous circulations
   B pulmonary and systemic circulations
   C arterial and capillary circulations
   D venous and capillary circulations

2 When the left atrium contracts, it pumps blood:
   A through an atrioventricular valve into the aorta
   B through the aortic valve into the aorta
   C through the aortic valve into the left ventricle
   D through an atrioventricular valve into the left ventricle

3 For the whole period of ventricular filling:
   A both atria are contracting
   B both ventricles are contracting
   C both atria are relaxing
   D both ventricles are relaxing

4 If the pressure in the left ventricle is higher than that in the left atrium, but lower than that in the aorta:
   A the atrioventricular valve is closed and the aortic valve is open
   B the atrioventricular valve is closed and the aortic valve is closed
   C the atrioventricular valve is open and the aortic valve is closed
   D the atrioventricular valve is open and the aortic valve is open

5 The ratio of the cardiac output of the left ventricle to that of the right ventricle is:
   A 2:1
   B 1:2
   C 1:1
   D none of the above

6 Cardiac output is equal to:
   A stroke volume of left ventricle × heart rate
   B stroke volume of right ventricle × heart rate
   C either of the above
   D neither of the above

7 In early ventricular systole, blood is not ejected from the ventricles because:
   A the pressure in the atria is higher than that in the ventricles
   B the pressure in the atria is lower than that in the ventricles
   C the pressure in the aorta is higher than that in the ventricles
   D the pressure in the aorta is lower than that in the ventricles

8 Arteries carry:
   A blood under high pressure towards the heart
   B blood under low pressure towards the heart
   C blood under low pressure away from the heart
   D blood under high pressure away from the heart

9 Impulses from the SA node are held up at the AV node. This is important because:
   A it allows atrial systole to be completed before ventricular systole commences
   B it allows ventricular systole to be completed before atrial systole commences
   C both ventricles do not enter systole at the same time
   D both atria do not enter diastole at the same time

10 Impulses generated by the SA node pass to the ventricles in the sequence:
   A atrial Purkyne tissue, bundle of His, AV node
   B atrial Purkyne tissue, AV node, bundle of His
   C bundle of His, atrial Purkyne tissue, AV node
   D bundle of His, AV node, atrial Purkyne tissue

Examination-style

1 The diagram shows a human heart, seen from the front.

(a) Name the structures labelled A, B and C. *(3 marks)*

(b) Some babies are born with a hole in the septum. How might this affect the composition of blood entering the systemic circulation? Explain your answer. *(4 marks)*

(c) Describe the conditions under which the valves labelled X and Y would open. *(2 marks)*

Total: 9 marks

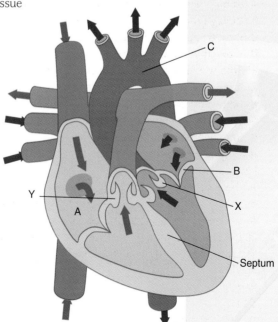

2 The graph shows the pressure changes in the left atrium, left ventricle and aorta during one cardiac cycle.

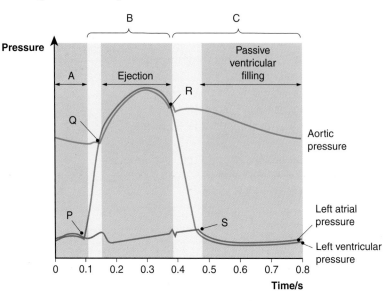

(a) Name the period of the cardiac cycle labelled B.
Give reasons from the graph for your choice. *(2 marks)*

(b) What happens at the points in the cycle labelled:
(i) P
(ii) Q
Explain your answers. *(4 marks)*

(c) Sketch a line on the graph that represents the changes in pressure in the *right* ventricle. *(1 mark)*

Total: 7 marks

3 (a) The diagram shows the main stages of the cardiac cycle.

Use numbers from the diagram to answer the following questions. Some questions may require more than one number to provide a full answer.

Which of the numbered stages shows:
(i) atrial systole
(ii) ventricular diastole
(iii) blood being pumped to the lungs *(3 marks)*

(b) (i) What is cardiac output? *(2 marks)*
(ii) Give *two* conditions under which cardiac output may be increased. *(2 marks)*

Total: 7 marks

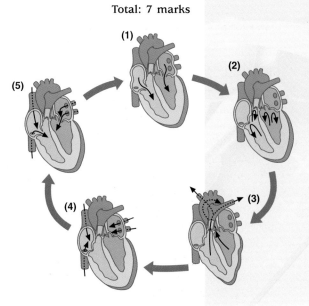

4 The graph shows the changes in volume of the left ventricle of a resting person over a period of time.

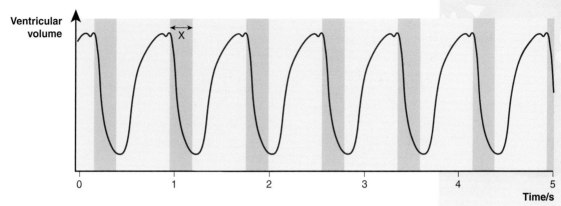

(a) (i) Describe and explain the main events of the cardiac cycle that occur during the period marked X. *(3 marks)*

(ii) Use the graph to calculate the heart rate of the person at rest. Show how you arrived at your answer. *(2 marks)*

(b) (i) Sketch a graph to show the changes in the volume of the left ventricle during a period of vigorous exercise. *(2 marks)*

(ii) Explain the reasons for the differences between the graph you have drawn and the graph above. *(3 marks)*

Total: 10 marks

5 Aortic and atrioventricular valves can become damaged, as a result of infection or due to an inherited condition. Much research has been carried out into the development of replacement valves. One artificial valve that has been in use for many years is the Starr–Edwards valve. In this valve, blood moves a small metal ball within a cage to alternately open and close the valve. The photograph shows a Starr–Edwards valve. The diagram shows the locations where such valves may be used in the heart.

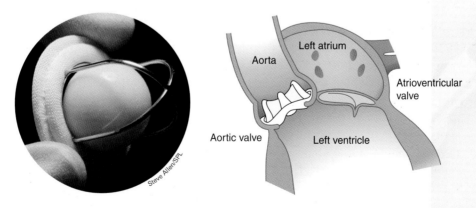

(a) Copy the diagram, omitting the atrioventricular and aortic valves. Instead, show the orientation you would expect of a Starr–Edwards valve in each location. *(2 marks)*

(b) Research is continuing into improving the materials used to make the cage and ball of the Starr–Edwards valve. Suggest one property of materials used for the ball and one property of materials used for the cage that might be researched. In each case, give a reason for your answer. (Note: mass and size are *not* properties of materials.) *(4 marks)*

(c) In one study, surgeons followed the progress of eight children born with a defect in the left atrioventricular valve. All had a Starr–Edwards valve implanted when they were less than 5 years of age. The surgeons reported the following:

- all patients developed normally to adult size
- all valves performed well for 160 patient-years
- one patient required a routine re-valving

(i) What was the average length of the follow-up period? Explain how you arrived at your answer. *(2 marks)*

(ii) Suggest why one patient might have needed a 'routine re-valving'. *(2 marks)*

(iii) Explain the importance of reporting that all patients developed to a normal adult size. *(3 marks)*

(d) Explain *one* reason why the results of this study should be treated with caution. *(2 marks)*

Total: 15 marks

Chapter 6

Breathing oxygenates the blood to allow food to be respired

This chapter covers:
- the structure of the human gas exchange system
- the mechanism of breathing
- gas exchange in the alveoli
  - the concept of partial pressure
  - the difference in composition of atmospheric and exhaled air

Breathing is necessary for gas exchange and energy release. On inhalation, atmospheric air (a mixture of gases) flows into the alveoli in the lungs. Here, gas exchange takes place and a different mixture of gases is exhaled. The main differences in composition between inhaled and exhaled air are due to the exchange of oxygen and carbon dioxide in the alveoli. The oxygen absorbed into the red blood cells is transported to all living cells to allow the release of energy from organic molecules obtained from food in aerobic respiration. The carbon dioxide produced by this process is carried in the blood plasma to the lungs, to be exhaled. The structure of the human gas exchange system is shown in Figure 6.1.

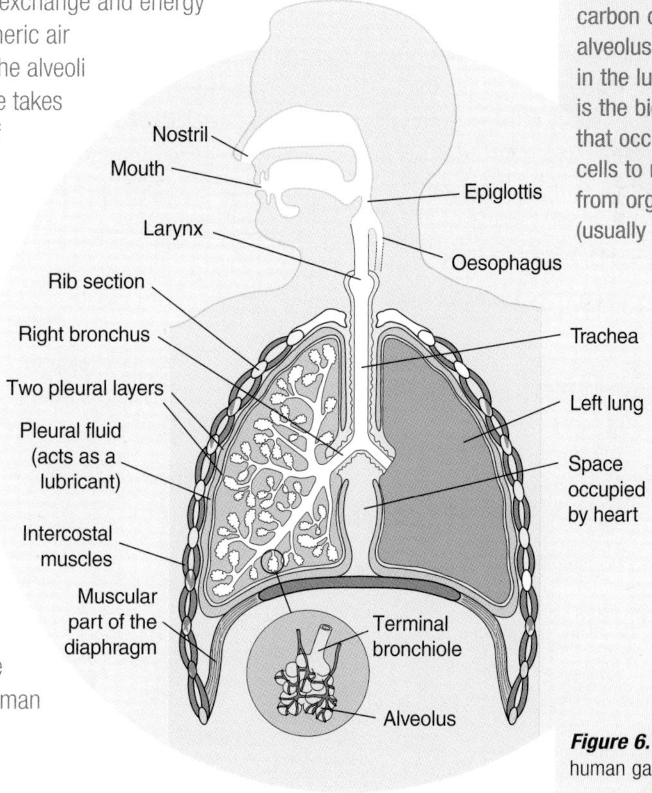

◀ Do not confuse breathing, gas exchange and respiration. Breathing is the movement of air into and out of the lungs. Gas exchange is the diffusion of oxygen and carbon dioxide across the alveolus and capillary walls in the lungs. Respiration is the biochemical process that occurs in all living cells to release energy from organic molecules (usually glucose).

Nostril
Mouth
Epiglottis
Larynx
Oesophagus
Rib section
Trachea
Right bronchus
Left lung
Two pleural layers
Pleural fluid (acts as a lubricant)
Space occupied by heart
Intercostal muscles
Muscular part of the diaphragm
Terminal bronchiole
Alveolus

Figure 6.1 Structure of the human gas exchange system

How do we breathe?

The breathing system moves inhaled air through the various passages into the alveoli, where gas exchange takes place. The air in the alveoli is then exhaled. Air is a mixture of gases and, like any gas, it moves because of pressure differences. Gases move from a region of high pressure to a region of low pressure. To move air in both directions, the relationship between the pressure in the lungs and the pressure in the atmosphere must change.

To bring air into the lungs (**inhalation**), the pressure must be lower in the lungs than in the atmosphere. To move air out again (**exhalation**), the pressure must be higher in the lungs than in the atmosphere. Breathing movements create these pressure differences.

The lungs are located in the **thorax**. This region of the body is separated from the **abdomen** by the **diaphragm**. Viewed from above (or below), the diaphragm is seen to have two main regions:

- an external muscular region, the outer edge of which is attached to the body wall
- a central fibrous region, made from tough connective tissue, which cannot contract or relax, be stretched or be compressed

Contraction of the muscular region of the diaphragm pulls the fibrous region downwards from its normal dome-shaped position. This flattens the diaphragm and enlarges the thoracic cavity.

The lungs are protected by the ribcage. There are two sets of muscles between each pair of ribs — the **external intercostal muscles** and the **internal intercostal muscles**. These muscles are attached in different ways. Therefore, their contractions produce different effects. Contraction of the external intercostal muscles lifts the ribs upwards and outwards; contraction of the internal intercostals muscles pull the ribs downwards and inwards.

Inhalation

The mechanism of inhalation is shown in the flowchart below:

When you turn on a gas tap to light a Bunsen burner, gas flows from the region of high pressure in the gas pipe to the lower pressure area in the Bunsen tubing.

Breathing movements alter the pressure between the lungs and the pleural layers that surround the lungs.

In humans, the internal intercostal muscles play little part in breathing movements at rest. We stand upright on two legs, so once the external intercostal muscles stop contracting, the ribs fall back downwards and inwards under gravity. In four-legged mammals both sets of muscles are equally important because gravity neither assists nor hinders breathing movements.

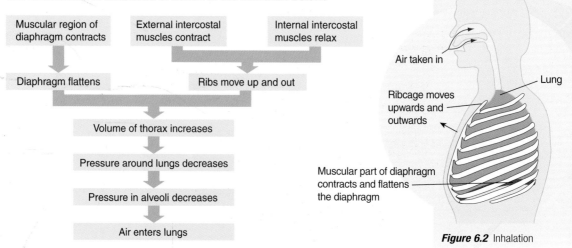

Figure 6.2 Inhalation

Exhalation

The mechanism of exhalation is shown in the flowchart below:

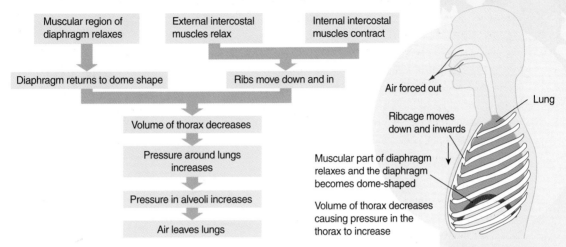

```
Muscular region of          External intercostal        Internal intercostal
diaphragm relaxes           muscles relax               muscles contract
        │                                                       │
        ▼                               ▼                       ▼
Diaphragm returns to dome shape          Ribs move down and in
        │                                       │
        └───────────────┬───────────────────────┘
                        ▼
              Volume of thorax decreases
                        ▼
              Pressure around lungs
                    increases
                        ▼
              Pressure in alveoli increases
                        ▼
                Air leaves lungs
```

Air forced out

Lung

Ribcage moves
down and inwards

Muscular part of diaphragm
relaxes and the diaphragm
becomes dome-shaped

Volume of thorax decreases
causing pressure in the
thorax to increase

Figure 6.3 Exhalation

What happens in the alveoli?

Moving air into the alveoli

Breathing movements move air in and out of the lungs through the airways that eventually lead to the alveoli. Air is drawn into the nose and/or mouth, through the pharynx (throat), down the trachea and along a bronchus into each lung. From here it passes along ever-finer bronchi and bronchioles until it reaches a terminal bronchiole, which leads to several alveoli via alveolar ducts.

The function of the airways is to allow the passage of air into and out of the alveoli, where gas exchange takes place.

The relationship between the terminal bronchiole, alveoli and capillaries is illustrated in Figure 6.4.

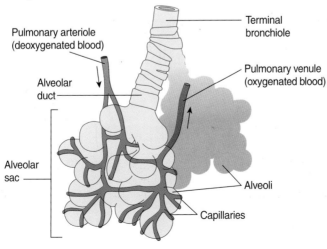

Pulmonary arteriole
(deoxygenated blood)

Alveolar
duct

Alveolar
sac

Terminal
bronchiole

Pulmonary venule
(oxygenated blood)

Alveoli

Capillaries

Figure 6.4 Alveoli are found at the end of terminal bronchioles

Monitoring breathing movements

The rate and depth of breathing movements can be monitored using a spirometer. The way that a spirometer works is shown in the diagram below.

When the subject breathes, pressure changes in the air in the spirometer produce movements of a recording pen. This produces a trace on a drum revolving at constant speed. The trace that is produced is called a spirogram. It shows both the depth and the rate of breathing movements. The spirogram below shows the breathing movements of a person initially at rest and then becoming more active.

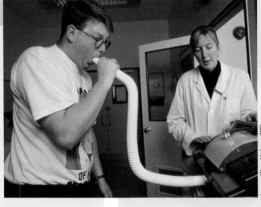

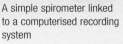

A simple spirometer linked to a computerised recording system

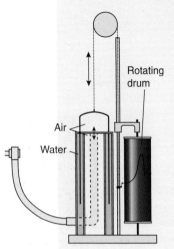

Rotating drum

Air

Water

At rest, we inhale around 500 cm³ of air with each breath and exhale the same amount. This is known as the tidal volume. We could inhale more — we can all *choose* to inhale deeply. This extra potential to inhale is called the inspiratory reserve volume — about 3000 cm³ in an average adult. Similarly, without inhaling any more deeply, we can decide to exhale more forcefully. This extra exhalation is the expiratory reserve volume — about 1000 cm³ in an average adult. So, over and above the tidal volume, there is room for another 3000 cm³ air and an extra 1000 cm³ can be forced out. Therefore, the total amount of air that can be brought in and out of the lungs when breathing as forcefully as possible is about 4500 cm³. This is the vital capacity. However, we can never completely empty our lungs of air. The air left after the most forceful exhalation is the residual volume — about 1200 cm³.

total lung capacity = vital capacity + residual volume = 5700 cm³

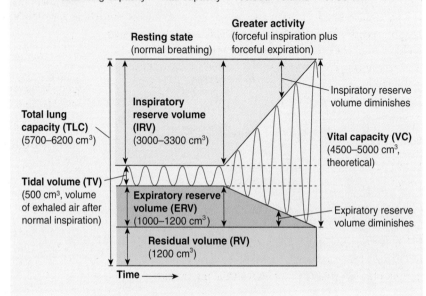

Resting state (normal breathing)

Greater activity (forceful inspiration plus forceful expiration)

Inspiratory reserve volume diminishes

Inspiratory reserve volume (IRV) (3000–3300 cm³)

Total lung capacity (TLC) (5700–6200 cm³)

Vital capacity (VC) (4500–5000 cm³, theoretical)

Tidal volume (TV) (500 cm³, volume of exhaled air after normal inspiration)

Expiratory reserve volume (ERV) (1000–1200 cm³)

Expiratory reserve volume diminishes

Residual volume (RV) (1200 cm³)

Time →

As a consequence of being surrounded by blood, alveoli are lined with fluid. This fluid contains a surfactant that is produced and secreted by specialised epithelial cells in the wall of each alveolus (Figure 6.5). The surfactant reduces the surface tension of the fluid, which prevents the walls of the alveoli from collapsing and sticking shut during breathing.

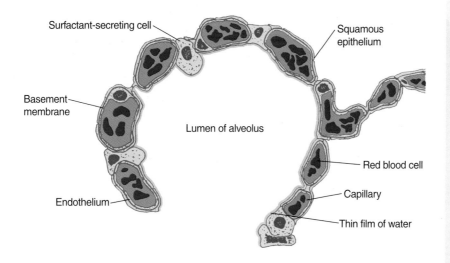

Figure 6.5 Section through an alveolus

What is the partial pressure of a gas?

The pressure exerted by the air is caused by molecules of the gases that make up the air colliding with, and exerting a force on, the surface of an object. The molecules of the various gases produce an almost identical force when they collide with the surface. So the total atmospheric pressure is the sum of the pressures caused by the molecules of all the different gases that make up the air. As 78% of the molecules in the air are nitrogen molecules, they account for 78% of the total pressure — this is the **partial pressure** of nitrogen. Twenty-one per cent of the air is oxygen, so oxygen accounts for 21% of the total pressure. Other gases (including carbon dioxide and water vapour) account for the remaining 1%. The partial pressure of oxygen is written as pO_2 and that of carbon dioxide as pCO_2. At sea level, atmospheric pressure (the sum of the partial pressures of all the gases in the atmosphere) is 100 kPa. Therefore, the partial pressure of oxygen is 21 kPa (21% of 100 kPa).

When you open a fizzy drink, it comes into contact with air that has a much lower partial pressure of carbon dioxide than that in the liquid. Carbon dioxide diffuses from the high partial pressure to the low partial pressure — quickly! The drink fizzes as carbon dioxide is lost.

$$\text{total atmospheric pressure} = pN_2 + pO_2 + pCO_2 + p\text{others}$$

$$100\,\text{kPa} = 78\,\text{kPa} + 21\,\text{kPa} + 0.03\,\text{kPa} + 0.97\,\text{kPa}$$
$$(100\%) \quad (78\%) \quad (21\%) \quad (0.03\%) \quad (0.97\%)$$

When a gas dissolves in a liquid, its molecules continue to exert a pressure and contribute to the total pressure of the liquid. The amount of a gas that can dissolve in a liquid is dependent on the partial pressure of the gas in the air around the liquid. Molecules of the gas diffuse into (or out of) the liquid until the two partial pressures are the same.

Figure 6.6 Partial pressures of atmospheric gases

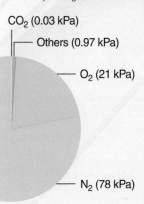

CO$_2$ (0.03 kPa)

Others (0.97 kPa)

O$_2$ (21 kPa)

N$_2$ (78 kPa)

Table 6.1 shows the partial pressures of oxygen and carbon dioxide in the atmosphere, in air in the alveoli and in blood plasma (in the pulmonary artery and in the pulmonary vein). Diffusion takes place until the partial pressures of the gases are the same in the alveolar air and the blood plasma.

Table 6.1

	Atmosphere/ kPa	Alveolar air/ kPa	Blood plasma in pulmonary artery/kPa	Blood plasma in pulmonary vein/kPa
Oxygen	21.00	11.20	5.30	11.20
Carbon dioxide	0.03	5.30	6.10	5.30

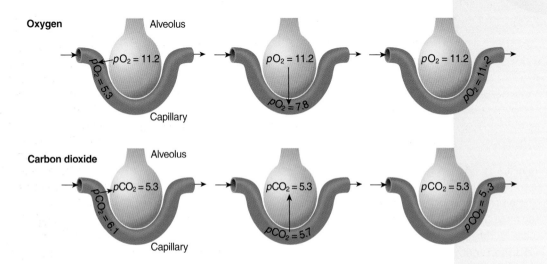

Figure 6.7 Exchange of O_2 and CO_2 between an alveolus and blood plasma

Partial pressure determines only how much oxygen and carbon dioxide are transported in simple solution in the blood plasma. However, most oxygen is transported combined with haemoglobin in the red blood cells. Only $0.3\,cm^3$ of oxygen is dissolved in $100\,cm^3$ plasma; the haemoglobin in the same volume of plasma carries $20\,cm^3$ oxygen. Gas exchange in the lungs saturates the plasma *and* the haemoglobin with oxygen.

Can we relate gas exchange in the lungs to Fick's law of diffusion?

Gas exchange in the alveoli occurs by passive diffusion. Therefore, the factors of Fick's law affect the overall efficiency of the process:

$$\text{rate of diffusion} \propto \frac{\text{surface area of exchange} \times \text{difference in concentration}}{\text{thickness of exchange surface}}$$

For efficient gas exchange, there must be a fast diffusion rate. This requires:
- a large surface area
- a large difference in concentration
- a short diffusion distance

The large surface area is provided, in part, by the 700 million alveoli present in the lungs of a human. Collectively, the alveoli have the area of a tennis court. However, it is the area over which exchange can actually take place that is important. This is represented by the total area of alveolar wall in contact with capillaries. So, having a vast number of capillaries is also important in providing a large exchange surface.

The difference in concentration is maintained by constant ventilation of the lungs and circulation of the blood. Ventilation continually replaces air in the alveoli with atmospheric air that has a high partial pressure of oxygen and a low partial pressure of carbon dioxide. Circulation removes newly oxygenated blood from the capillaries next to the alveoli and replaces it with deoxygenated blood. The partial pressure of oxygen in this blood is lower than that of the alveolar air; the partial pressure of carbon dioxide in this blood is higher than that of alveolar air.

If you look back to Table 6.1, you will see that the concentration gradient between the pulmonary artery and alveolar air is greater for oxygen than it is for carbon dioxide. Despite this, the volumes of the two gases exchanged are almost identical. How can this be? This is a consequence of the moisture lining the alveoli. Carbon dioxide is much more soluble in water than oxygen is and it dissolves quickly in the water that lines the alveoli. This allows a more rapid transfer across the alveolar walls.

The short diffusion distance is a consequence of:

- the extreme thinness of the alveolar wall and the capillary wall — both comprise only squamous epithelial tissue
- alveoli and capillaries being pressed closely together— the interstitial space (space between the two) is very small
- the shape of the red blood cell — most of the oxygen diffuses from the air and combines with haemoglobin in red blood cells, the flattened shape of which provides a large surface area and also means that the haemoglobin is close to the plasma membrane, reducing the diffusion distance

Squamous epithelial cells are the thinnest cells in our bodies. Most of the cell is thinner than its nucleus, which causes a 'bulge' in the cell. A typical squamous cell is less than 1 µm thick, whilst its nucleus is 6 µm thick. By comparison, the epithelial cells lining the gut are approximately ◀ 30 µm thick.

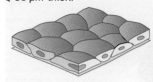

How does the composition of the air change while in the lungs?

As a result of the diffusion of gases in the alveoli, the composition of air that is exhaled is different from that which is inhaled and from that which is present in the alveoli (Table 6.2).

Table 6.2

	Inhaled air/%	Alveolar air/%	Exhaled air/%
Oxygen	21.00	11.20	16.00
Carbon dioxide	0.03	5.30	3.00

As you might expect, alveolar air and exhaled air both contain less oxygen and more carbon dioxide than inhaled air. But why are the two different? This is because exhaled air is a mixture of air from the alveoli and air in the bronchi and bronchioles that did not reach the alveoli. This air is not involved in gas exchange and, therefore, still has the same composition as atmospheric (inhaled) air. As a result, exhaled air has concentrations of oxygen and carbon dioxide that are intermediate between those of alveolar air and inhaled air.

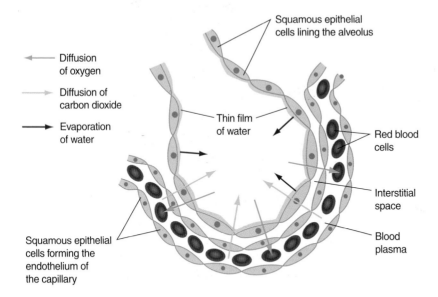

Diffusion of oxygen

Diffusion of carbon dioxide

Evaporation of water

Squamous epithelial cells lining the alveolus

Thin film of water

Red blood cells

Interstitial space

Blood plasma

Squamous epithelial cells forming the endothelium of the capillary

Figure 6.8 The diffusion pathway between a capillary and an alveolus

How much air do we breathe?

At rest we have a breathing rate of about 15 breathing movements (inhalations and exhalations) per minute and a tidal volume of around $500\,cm^3$ ($0.5\,dm^3$). This means that a total of $7500\,cm^3$ ($7.5\,dm^3$) is breathed in per minute. This is called the minute volume or the **pulmonary ventilation rate**.

pulmonary ventilation rate = tidal volume × rate of breathing

However, on exercising, an increased pulmonary ventilation rate is needed to supply the extra oxygen. Also, following exercise, the pulmonary ventilation rate is affected by high levels of lactate that accumulate in muscles as a result of anaerobic respiration. After exercise, lactate is removed by a process that needs oxygen. The pulmonary ventilation rate falls slowly as more and more lactate is oxidised.

Box 6.2 VO₂ max: a measure of athletic fitness

Athletic performance is largely dependent on how much oxygen can be supplied to the skeletal muscles.

This depends on how much blood can be delivered to the muscles and how effectively the blood is oxygenated during the exercise.

The units of VO₂ max are cm^3 per minute per kilogram of body mass. One way to improve fitness is to lose weight. Does this work according to the VO₂ max formula?

VO$_2$ max is given by dividing the volume of oxygen used by the time it takes to perform the exercise and then dividing again by body mass. If you lose weight, you will divide by a smaller number and so increase your VO$_2$ max. You will also find it easier to move the new, lighter you, and complete the task in a shorter time, again improving your VO$_2$ max.

VO$_2$ max is determined experimentally by analysing the exhaled air of an athlete as he or she trains on a treadmill.

It can be determined more easily by an 'indirect' method. For this, warm up gently, walk one mile as quickly as you can (note the time taken) and take your pulse immediately after completing the exercise. Then, use the following formula to determine VO$_2$ max.

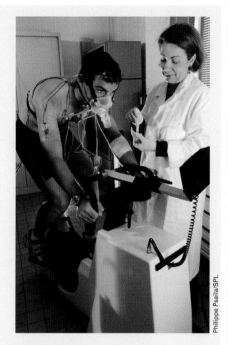

Philippe Psaila/SPL

$$VO_2 \text{ max } (cm^3 \, min^{-1}kg^{-1}) = 132.853 - (0.0769 \times \text{weight in pounds})$$
$$- (0.3877 \times \text{age in years}) + (6.3150 \times \text{sex}^*)$$
$$- (3.2649 \times \text{time (in minutes)}) - (0.1565 \times \text{final heart rate})$$

(*Male $= 1$, female $= 0$)

The following tables show VO$_2$ max values for men and women that represent a range of performance levels.

VO$_2$ values for men:

	20–29 (yrs)	30–39 (yrs)	40–49 (yrs)	50–59 (yrs)	60–69 (yrs)
Low	<38	<34	<30	<25	<21
Quite low	39–34	35–39	31–35	26–31	22–26
Average	44–51	40–47	36–43	32–39	27–35
High	52–56	48–51	44–47	40–43	36–39
Very high	>57	>52	>48	>44	>40

VO$_2$ values for women:

	20–29 (yrs)	30–39 (yrs)	40–49 (yrs)	50–65 (yrs)
Low	<28	<27	<25	<21
Quite low	29–34	28–33	26–31	22–28
Average	35–43	34–41	32–40	29–36
High	44–48	42–47	41–45	37–41
Very high	>49	>48	>46	>42

Summary

Structure

- The human breathing system comprises:
  - the airways in the lungs (bronchi, bronchioles, alveoli) and those outside the lungs (trachea, larynx, nasal and mouth cavities)
  - the structures that surround the lungs and assist in the mechanisms of breathing movements (ribs, intercostal muscles and diaphragm)

Mechanism

- Inhalation and exhalation occur because breathing movements create pressure differences between the atmosphere and the air in the lungs.
- Inhalation is brought about because the contraction of the external intercostal muscles and contraction of the diaphragm muscle increase the volume of the thorax, which reduces pressure in the thorax.
- Exhalation is brought about because the contraction of the internal intercostal muscles and the relaxation of the diaphragm muscle decrease the volume of the thorax and this increases pressure in the thorax.
- Inhalation draws air though the nasal cavity and pharynx, along the trachea, bronchi and bronchioles, and into the alveoli, where gas exchange takes place. Exhalation forces air from the alveoli back along the same route.

Gas exchange

- In the alveoli, oxygen diffuses from the alveolar air into the red blood cells (where it combines with haemoglobin), and carbon dioxide diffuses from the blood plasma into the alveolar air.
- Diffusion in the alveoli is efficient (in terms of Fick's law) because:
  - the many alveoli provide a large surface area
  - continuous ventilation and circulation maintain a high concentration difference between alveolar air and the blood
  - the extremely thin walls of the alveoli and capillaries (consisting only of squamous epithelium on a basement membrane) provide a short diffusion distance
- Gas exchange is also more efficient because the alveoli are moist; this aids the diffusion of carbon dioxide in particular.
- The partial pressure of a gas is the contribution made by that gas to the total pressure of a system. It is a useful way of comparing concentrations of gases between a gaseous medium and a liquid medium.
- The volume of air moved in and out of the lungs with each breath is the tidal volume.
- At rest, the tidal volume is about 500 cm^3, but the inspiratory reserve volume and the expiratory reserve volume can increase this to about 4500 cm^3.
- There is about 1200 cm^3 air that is never exhaled, which forms the residual volume.

Control

● pulmonary ventilation rate = rate of breathing × tidal volume

Questions

Multiple-choice

1 Pulmonary ventilation rate is equal to:
  A breathing rate × residual volume
  B breathing rate × tidal volume
  C tidal volume ÷ breathing rate
  D residual volume ÷ breathing rate

2 Diffusion of oxygen and carbon dioxide in the alveoli is rapid because there is:
  A a large surface area, a low concentration difference and a short diffusion pathway
  B a small surface area, a high concentration difference and a short diffusion pathway
  C a large surface area, a high concentration difference and a long diffusion pathway
  D a large surface area, a high concentration difference and a short diffusion pathway

3 To bring about inhalation:
  A external intercostal muscles contract, internal intercostal muscles relax, diaphragm muscle contracts
  B external intercostal muscles contract, internal intercostal muscles contract, diaphragm muscle relaxes
  C external intercostal muscles relax, internal intercostal muscles relax, diaphragm muscle contracts
  D external intercostal muscles relax, internal intercostal muscles contract, diaphragm muscle relaxes

4 During gas exchange in the alveoli, oxygen must diffuse through:
  A the alveolar epithelium
  B the interstitial space
  C the membrane of a red blood cell
  D all of the above

5 The tidal volume is:
  A the total volume of air in the lungs
  B the volume inhaled added to the volume exhaled in one breath
  C the volume of air inhaled and then exhaled in one breath
  D the maximum volume of air that can be inhaled in one breath

6 The partial pressure of a gas is:
  A the proportion by volume of a gas in a mixture of gases
  B the part of the total pressure that a gas exerts due to its molecules colliding with each other
  C the part of the overall pressure of a mixture of gases that is due to that gas
  D the pressure of a gas when only some of its molecules exert a force

7 When air is exhaled, it:
   A diffuses out down a concentration gradient
   B is drawn out because of the higher pressure in the atmosphere
   C is forced out because of the higher pressure in the lungs
   D is drawn in because of the lower pressure in the lungs
8 Following exercise, the pulmonary ventilation rate falls only slowly because
   A the partial pressure of oxygen in the blood is high
   B the levels of lactate are still high
   C the levels of lactate are still low
   D the partial pressure of oxygen in the blood is low
9 Ventilating the lungs means:
   A inhaling
   B exhaling
   C inhaling and exhaling
   D exchanging gases
10 When compared with inhaled air, exhaled air:
   A has a higher concentration of oxygen, a lower concentration of carbon dioxide and is more moist
   B has a higher concentration of oxygen, a lower concentration of carbon dioxide and is drier
   C has a lower concentration of oxygen, a higher concentration of carbon dioxide and is more moist
   D has a lower concentration of oxygen, a higher concentration of carbon dioxide and is drier

Examination-style

1 (a) The diagram below shows the main components of the human breathing system.

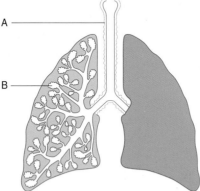

   (i) Name the structures labelled A and B. *(2 marks)*
   (ii) Explain why contraction of the external intercostal muscles and contraction of the diaphragm muscle cause air to be drawn into the lungs. *(3 marks)*
   (b) Use Fick's law to explain why gas exchange in the alveoli is efficient. *(4 marks)*

   Total: 9 marks

2 The diagram shows two spirometer traces taken from the same person at different times.

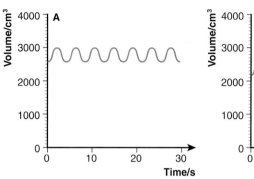
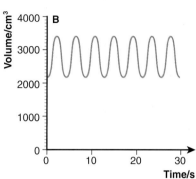

(a) (i) What is the tidal volume shown in trace A? *(1 mark)*
   (ii) From trace B, calculate the pulmonary ventilation rate.
      Show your working. *(2 marks)*
(b) Suggest reasons for the difference in breathing shown by
   the two traces. Explain your answer. *(4 marks)*

Total: 7 marks

3 (a) Explain what is meant by the terms:
   (i) tidal volume
   (ii) inspiratory reserve volume
   (iii) vital capacity *(3 marks)*
(b) When a person suffers a serious chest wound, a condition known as pneumothorax can result. In this condition, air rushes into the space between the two pleural layers surrounding one of the lungs.
   (i) What effect will the wound have on the pressure of the air
      between the two pleural layers? Explain your answer. *(2 marks)*
   (ii) Explain why the lung affected by pneumothorax can no
      longer be ventilated. *(3 marks)*

Total: 8 marks

4 The diagram shows two spirometer traces. Trace A shows a normal breathing pattern of a person at rest. Trace B shows an abnormal pattern of breathing at rest.

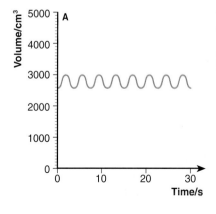
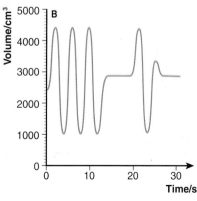

(a) Calculate the pulmonary ventilation rate shown by trace A. *(2 marks)*

(b) Describe two ways in which the pattern of breathing shown by trace B differs from that shown in trace A. *(2 marks)*

(c) The pattern of breathing shown in trace B is sometimes found in brain-damaged patients, where the nervous control of breathing fails and then begins again. Suggest how brain damage might produce this abnormal pattern of breathing. *(3 marks)*

Total: 7 marks

5 The graph below shows the atmospheric pressure at different altitudes.

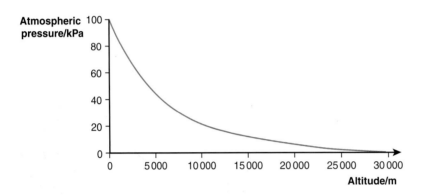

(a) Describe the change in atmospheric pressure between sea level and 10 000 m. *(2 marks)*

(b) Assuming that oxygen makes up 21% of the atmosphere at all altitudes, calculate the partial pressure of oxygen at:

(i) sea level *(1 mark)*

(ii) an altitude of 5000 m *(1 mark)*

(c) Gas exchange takes place in the alveoli. Most of the oxygen absorbed diffuses into red blood cells and combines with haemoglobin to form oxyhaemoglobin.

(i) Explain why oxygen combining with haemoglobin does not affect the concentration gradient of oxygen between the alveolar air and the blood. *(2 marks)*

(ii) Suggest why people who live at high altitudes have more red blood cells per cm^3 blood than people who live at sea level. *(3 marks)*

(iii) Use Fick's law to explain the benefit, to a person who has just reached a high altitude, of breathing more quickly. *(3 marks)*

(d) Explain to a person not adapted to living at such altitudes, why exercise is more difficult at higher altitudes. *(3 marks)*

Total: 15 marks

Chapter 7

Disease: when it all goes wrong

This chapter covers:

- the concept of disease
- ways in which disease can be caused
- the nature of pathogens, including:
  - bacteria
  - viruses
  - fungi
- the causative organism, transmission, symptoms and treatment of cholera
- the course of infection, transmission, symptoms and treatment of pulmonary tuberculosis (TB)
- the biological basis of heart disease and cancer
- the biological basis of fibrosis, asthma and emphysema

Complete physical, mental and social well-being makes a healthy person

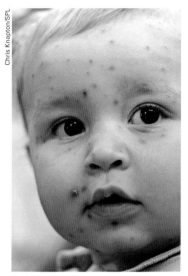

A child with chickenpox

The World Health Organization (WHO) defines health as 'a state of complete physical, mental and social well-being'.

Disease is less easy to define. It does not just mean the absence of perfect health. We may be less fit than we should be, or feeling depressed at the thought of revising for an examination, but that does not necessarily mean that we have a disease.

A useful definition of disease might be 'a condition with a specific cause in which part or all of the body is made to function in a non-normal and less efficient manner'. This definition could include diseases of all organisms — including plants. It could also include physical, mental and social aspects of disease in humans.

Chickenpox produces a fever and makes a person feel drowsy, itchy and generally unwell. Most people suffer no major complications, but occasionally the disease can be fatal.

What causes disease?

When we think of disease, we usually think of infectious disease — a disease caused by another organism that gains entry to the body. However, there are a number of ways in which the body can be made to function in a 'non-normal and less efficient manner'.

The causes of disease include:

- pathogenic organisms (bacteria, viruses, fungi and protoctistans). Diseases caused in this way are known as **infectious diseases**. Infectious diseases that can be transferred from one person to another are called **communicable diseases**. Cholera and pulmonary tuberculosis are both infectious diseases, but cholera is not a communicable disease whereas pulmonary tuberculosis is.
- a person's lifestyle and working conditions. These may result in **human-induced diseases**. Examples include many cancers, some forms of heart disease and fibrosis.
- degenerative processes. These are often the result of ageing. Arthritis and atherosclerosis are examples of **degenerative diseases**.
- our genes. Haemophilia and sickle-cell disease are examples of **genetic diseases**.
- nutrient deficiency. **Deficiency diseases** include scurvy (caused by a lack of vitamin C in the diet) and kwashiorkor (caused by a lack of protein in the diet).
- social activities. **Social diseases**, including alcoholism and drug addiction, may result in dependency on the drug, isolation, clinical depression and various levels of anti-social behaviour.

◀ Eating too much food can result in obesity, which is itself regarded as a disease condition and which can also lead to other diseases, such as coronary heart disease.

A drug addict. This person is not in a state of complete physical, mental and social well-being. His body is functioning in a less efficient and non-normal manner. His condition has a clearly defined cause.

Faye Norman/SPL

Box 7.1 Categorising diseases

In some cases, it is an oversimplification to place a disease in a single category. For example, atherosclerosis (the laying down of fatty substances in arteries) increases with age, so it can be classified as a degenerative disease. However, this process is also influenced by diet; the more saturated fat we eat, the more fatty substances are laid down in our arteries. There is a separate genetic component; some people have an increased risk of developing this disease. Stress and high blood pressure also increase the rate at which atherosclerosis develops. Both of these can be the result of lifestyle. Atherosclerosis does not fit neatly into one category. It is best to consider such conditions as **multifactorial**.

What sort of organisms infect us?

Most organisms that infect humans and cause disease are microorganisms. Organisms that cause disease are called **pathogens**. The process by which a pathogen enters and becomes established in an organism to cause disease is called **infection**.

Most infectious diseases in humans are caused by either bacteria or viruses, although a few are caused by fungi. Bacteria are single-celled organisms; viruses are acellular (not made of cells). Fungi can be single-celled (e.g. yeast) or multi-cellular.

The photographs below show some pathogenic microorganisms.

Don't refer to the disease itself as an infection. Disease is a condition; ◀ infection is a process.

Pathogenic microorganisms:
(a) *Salmonella* (a bacterium)
(b) adenovirus particles
(c) *Candida albicans* (a fungus)

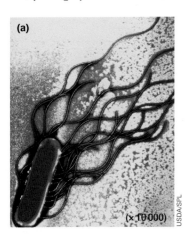

(a) (×10 000) USDA/SPL

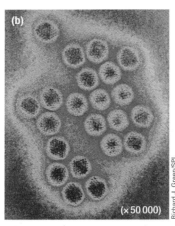

(b) (×50 000) Richard J. Green/SPL

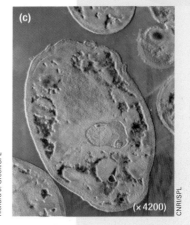

(c) (×4200) CNRI/SPL

Figure 7.1 shows the structure of the different types of pathogens.

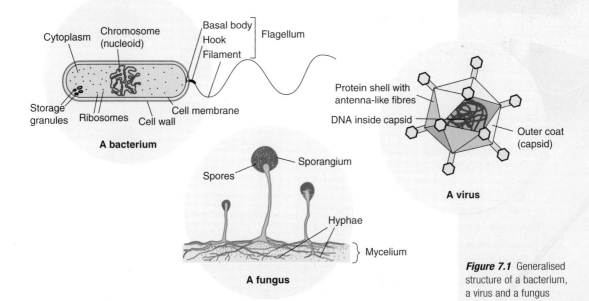

Figure 7.1 Generalised structure of a bacterium, a virus and a fungus

Table 7.1 summarises how the different types of pathogens cause disease.

Table 7.1

Type of microorganism	How the microorganism causes disease	Example of disease caused
Bacterium	Bacteria release toxins as they multiply. These toxins affect cells in the region of the infection and sometimes also in other regions of the body. Bacterial diseases can be treated with antibiotics because each bacterium is a cell with its own 'metabolic systems' and is capable of cell division. Antibiotics can enter the bacterial cells and disrupt these processes.	Pulmonary tuberculosis (TB); pneumonia; cholera
Virus	Viruses enter living cells and disrupt their metabolic systems. The genetic material of the virus becomes incorporated into that of the cell and instructs the cell to produce more viruses. Diseases caused by viruses cannot be treated with antibiotics because viruses are not true cells and are only active inside human cells, which antibiotics cannot enter.	Influenza ('flu); AIDS; measles; common cold
Fungus	When fungi grow in or on living organisms, their hyphae (see Chapter 15, page 267) secrete enzymes. These digest substances in the tissues; the products of digestion are absorbed. Growth of hyphae physically damages the tissue. Some fungi also secrete toxins. Others can cause allergic reactions.	Athlete's foot; farmer's lung

How does the cholera bacterium cause disease?

Cholera is an infectious disease of the intestines. It is not normally a communicable disease; there are very few recorded cases of cholera being transmitted from person to person. Cholera is caused by the bacterium *Vibrio cholerae*. This bacterium gains entry to the body in polluted drinking water and, occasionally, in food. In areas where sanitation is poor, cholera is often common. It frequently appears in an area following a natural disaster, or after war, when treatment of drinking water is affected. However, cholera is not always a killer disease. One estimate suggests that 80% of people infected with the cholera bacterium have no symptoms at all. Many more have only mild symptoms.

Like all pathogenic bacteria, *Vibrio cholerae* produces toxins. These toxins affect the permeability of the epithelial cells of the small intestine, allowing large amounts of water and salts to be lost from the cells into the lumen of the small intestine. If this continues, it causes dehydration, which can be fatal if not treated properly.

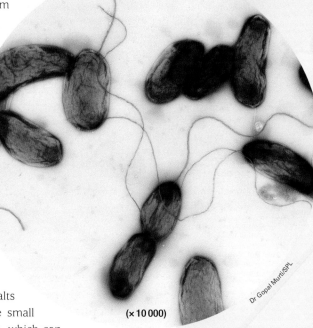

Vibrio cholerae: just two strains of this bacterium are responsible for all the current outbreaks of cholera

Dr Gopal Murti/SPL

(×10 000)

The cholera toxin alters transport proteins in the plasma membrane of the epithelial cells, causing them to be permanently 'open'. This allows ions to flow out of the cell. This sets up a water potential gradient that causes water to follow by osmosis. This water must ultimately come from the blood. As more and more water is lost from the blood, the circulatory system can collapse, resulting in death.

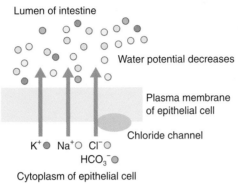

Lumen of intestine

Water potential decreases

Plasma membrane of epithelial cell

Chloride channel

K⁺● Na⁺○ Cl⁻○

HCO₃⁻○

Cytoplasm of epithelial cell

Figure 7.2 How the cholera toxin works

Box 7.2 How cholera spreads across the globe

There have been seven known cholera **pandemics**. The first known pandemic occurred in 1816; the seventh pandemic is ongoing. Until recently, it was thought that the most likely way for cholera to be carried to new areas was by infected, but unaffected, 'carriers'. Although carriers could not transmit the disease, they could bring the bacterium to new areas where, if sanitation was poor, the bacterium could infect other people. More recently, however, it has been shown that cholera bacteria also infect tiny marine crustaceans called copepods. These copepods migrate to follow the plankton (their food) in the oceans. While in the copepods, the cholera bacteria are largely dormant, but suitable environmental conditions of temperature and salinity seem to trigger their return to a more active state.

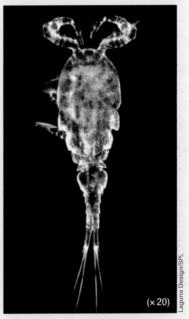

(× 20)

Laguna Design/SPL

Light micrograph of a copepod

◀ A *pandemic* occurs when a disease that is normally localised (found in one area) becomes global. In its localised condition, the disease is said to be *endemic* to that area. A non-global spread from that area is an *epidemic*.

Although cholera can be a killer, the WHO states that less than 1% of even serious cases would prove fatal if properly treated. So what is proper treatment? The main treatment is to administer oral rehydration therapy (see Chapter 4, page 76). This prevents the dehydration that can lead to death. It does not cure the disease, but, by maintaining a proper fluid balance, most affected people are capable of making an effective immune response and a full recovery.

Box 7.3 The cholera story: the work of one scientist makes it possible for others to take it further

Cholera had been known for many hundreds of years and, like many other diseases, it was believed to be transmitted by 'malodorous air' or 'miasma'. In 1849, John Snow proposed that cholera was, in fact, transmitted by water. In 1854, a new outbreak of cholera began in London. At the time, two main water companies supplied central London with water. The Lambeth Company drew water from the river Thames upstream of the city; the Southwark and Vauxhall Company drew it downstream of the city — after it had been polluted by sewage. Snow realised that he had all the makings of what he called a 'grand experiment'. He was able to record the number of deaths from cholera in all areas of London and relate this to the water supply. His results are shown in Table 7.2.

Water supply	Number of houses	Deaths from cholera	Deaths per 10 000 houses
Southwark and Vauxhall Company	40 046	1263	315
Lambeth Company	26 107	98	37
Rest of London	256 423	1422	59

Table 7.2

The highest incidence of deaths from cholera occurred in the areas supplied by the Southwark and Vauxhall Company. Snow suggested that the polluted water supplied by this company was the cause of the outbreak. However, although there was clearly a strong link, the information was only correlational; it did not establish cause and effect.

Snow was able to go a little further when he found that there was a high concentration of deaths from cholera near to one particular water pump in Broad Street (Figure 7.3)

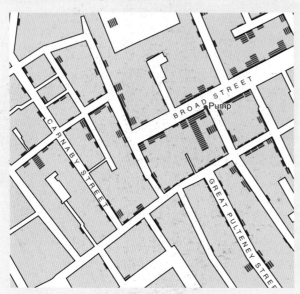

Each bar on the map represents a single death from cholera. Notice the high concentration near the Broad Street pump.

◀ Although Snow called it an experiment, it wasn't an experiment in the true sense, as it did not isolate an independent variable (cause) and a dependent variable (effect), with all other variables controlled. It is best described as a 'natural experiment' because there was a natural variation in the water supply. The large number of people involved made it likely that the data would be representative.

◀ At that time, most people still obtained their water from pumps in the street.

Figure 7.3 Part of John Snow's map of the 1854 cholera outbreak in London

By systematic observation and recording, Snow had been able to provide evidence in favour of his hypothesis concerning the transmission of cholera. He hypothesised that if the water pump was put out of use, the number of case would fall. He suggested that the water company remove the handle on the water pump. The number of new cases in the area quickly fell and the outbreak was contained.

Snow had provided strong evidence for a link between polluted water and the transmission of cholera, particularly by his analysis of the data in the Broad Street outbreak. However, he had not quite proved cause and effect. That was left to other scientists. At the same time as Snow was working in London, an Italian scientist, Filippo Pacini, isolated the bacterium *Vibrio cholerae*, but his work went unnoticed. In 1883, Robert Koch, unaware of Pacini's work, repeated the isolation, and was able to establish that the bacterium caused the disease. It also led him to develop the now famous '**Koch's postulates**'.

◀ In recognition of Pacini's work, the bacterium was officially renamed in 1965 as *Vibrio cholerae Pacini 1854*.

They state that:

- the microorganism must *always* be present when the disease is present, and should *not* be present if the disease is not present
- the microorganism can be isolated from an infected person and then grown in culture
- introducing such cultured microorganisms into a healthy host should result in the disease developing
- it should then be possible to isolate the microorganism from this newly diseased host and grow it in culture

The first postulate establishes a link between the microorganism and the disease. The following three postulates prove that the metabolism of a specific living microorganism, when transferred into a healthy host, causes the disease.

What causes pulmonary tuberculosis?

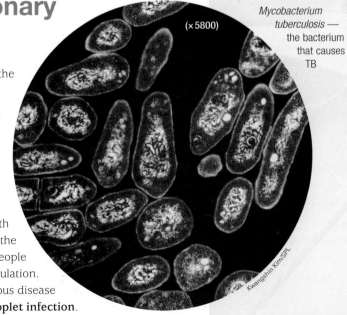

(× 5800)

Mycobacterium tuberculosis — the bacterium that causes TB

Kwangshin Kim/SPL

Pulmonary tuberculosis (TB) is caused by the bacterium *Mycobacterium tuberculosis*.

Robert Koch was the first to isolate *Mycobacterium tuberculosis*. The bacterium most commonly infects and affects the lungs, but other organs, such as the kidneys, the central nervous system, bones and skin, can also be affected.

Tuberculosis is the leading cause of death from a bacterial infectious disease in the world. The disease affects 1.7 billion people every year — one-third of the world's population. TB is a contagious (communicable) infectious disease that is spread from person to person by **droplet infection**.

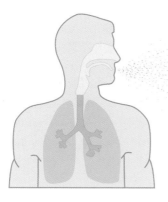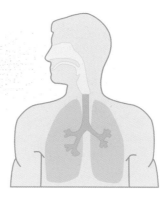

Figure 7.4 TB bacteria are spread in droplets. An infected person may cough or sneeze, releasing these droplets to be breathed in by another person.

The disease only begins to develop if the droplets reach the alveoli. It then progresses through the following stages:

- The bacteria are engulfed by macrophages (similar to phagocytic white blood cells), but not destroyed. They multiply inside the macrophages and eventually cause the macrophages to burst, releasing the bacteria.
- After about 21 days, T-lymphocytes (see Chapter 8, pages 142–43) begin to arrive at the site and activate the macrophages so that they can destroy the bacteria.
- If the bacteria are not eliminated by the immune response, **tubercles** begin to form. These are lumps with a semi-solid centre. The bacteria can survive in the tubercle, but not multiply. However, there could be many inactivated macrophages surrounding the tubercle and the bacteria multiply in these, causing the tubercle to enlarge.
- The centres of the tubercles liquefy and the bacteria multiply rapidly. As the tubercle grows, it may invade a bronchus (allowing the bacteria to spread to other parts of the lungs) or an artery (allowing it to spread to other parts of the body and create secondary infections). The rapid multiplication of the bacteria at this stage results in the formation of cavities in the lung. These cavities gave the disease its common name of 'consumption'.

◀ In developed countries, a good diet and good living conditions allow the body to mount an effective immune response. Only about 3–4% of people develop the active disease following an initial infection with *Mycobacterium tuberculosis*. The situation in developing countries is very different.

Symptoms of TB include:
- chest pain and coughing up blood
- a cough that lasts for more than three weeks
- chills and fever
- night sweats
- loss of appetite and loss of weight
- fatigue

People most at risk of infection include those who:
- are in regular contact with large numbers of people (e.g. in nursing homes, prisons and schools)
- have an inadequate diet
- inject drugs
- drink alcohol to excess
- are infected with HIV

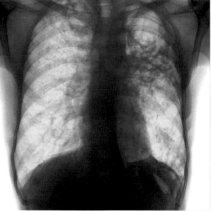

Gusto Images/SPL

A chest X-ray of a male patient with pulmonary tuberculosis. The affected areas of the lungs are the grainy dark patches at upper right.

Box 7.4 TB and AIDS

Ten per cent of all HIV-positive individuals are infected with the TB bacterium. This is 400 times the rate for the general population and makes HIV infection the main predisposing factor for TB. The map below shows the worldwide distribution of TB; the African continent has the highest incidence. This coincides with the incidence of AIDS, which is also highest in Africa.

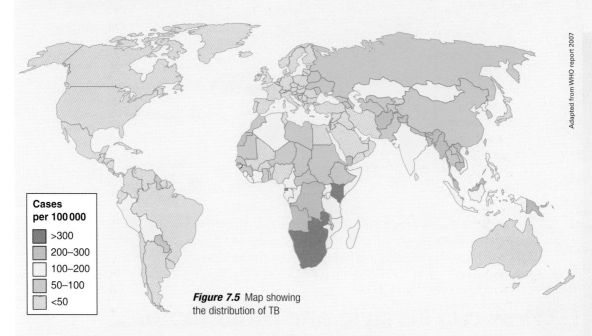

Adapted from WHO report 2007

Cases per 100 000

- >300
- 200–300
- 100–200
- 50–100
- <50

Figure 7.5 Map showing the distribution of TB

HIV infects the T-cells and, therefore, reduces the ability of the body to mount an effective immune response. As a result, the progression to active TB is much more common in AIDS sufferers.

Reducing the incidence of TB

This can be tackled on several fronts. Anything that reduces the risk of infection, or improves the body's ability to mount an effective immune response will reduce the incidence of the disease.

Treatment and vaccination

Effective treatment of TB became possible in 1946, when the antibiotic streptomycin was discovered. However, resistant forms of the bacterium developed and today TB is treated with a combination of four antibiotics. This increases the effectiveness of treating multi-drug resistant strains of the bacterium and is 95% effective.

The vaccine against TB is the BCG vaccine. This is named after the two Frenchmen (Calmette and Guerin) who developed the vaccine at the Pasteur Institute in 1921. It contains a live, but weakened form of the bacillus that causes TB in cattle. It is about 80% effective in children.

◀ BCG stands for **B**acille **C**almette-**G**uerin

Improvements in general hygiene and living conditions have reduced the transmission of the disease. Improvements in diet have meant that people who contract the disease are much more likely to produce an effective immune response. The graph below illustrates these points.

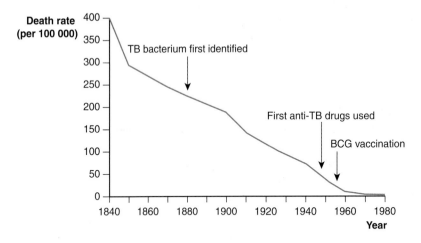

Figure 7.6 Death rate in England due to TB

In the UK, most cow's milk has been pasteurised or ultraheat-treated for many years. This, together with the regular testing of herds and the slaughter of any cattle with signs of infection, has reduced greatly the risk of transmission from infected cattle.

How can lifestyle and other factors lead to disease?

Obesity, the result of significant and prolonged over-eating, has been linked to many conditions. Stress is linked to heart disease and to some forms of cancer. Smoking is linked to heart disease, lung cancer and a number of other conditions. So just how do these factors make our bodies function in a 'less efficient and non-normal manner'?

The biological basis of heart disease

Heart disease has been on the increase for many years. Thanks to intensive research we now know a lot more about the causes of heart disease as well as how to avoid and treat the condition. A **heart attack** usually results from the sudden death of, or damage to, **cardiac muscle** in the wall of one of the ventricles of the heart — a **myocardial infarction**. Myocardial infarctions are usually the result of an interrupted blood supply to the cardiac muscle, which results in a lack of oxygen for aerobic respiration. Anaerobic respiration releases insufficient energy for the muscle cells, which fatigue and may die as a result. The most common cause of an interrupted blood supply is a blood clot or **thrombus** blocking one of the **coronary arteries**. The process of forming a blood clot is called **thrombosis**. If this happens in a coronary artery, it is a **coronary thrombosis**.

Box 7.5 Long-haul air flights

Long-haul air flights increase the risk of **deep vein thrombosis**. Many veins are located deep inside muscles and the contraction and relaxation of these muscles helps to move blood through the veins. When we sit still for long periods, blood in these veins moves less easily and tends to 'pool'. Under these conditions, a blood clot is much more likely to form, which may move and obstruct a coronary artery. Passengers on long-haul flights should get up and walk around periodically, to exercise these muscles and help move the blood in these veins.

The blood clot may not have formed in the coronary artery. It may have formed elsewhere in the body and then all, or part, of the clot may have become dislodged and travelled in the bloodstream to become lodged in another artery. It is then called an **embolus**. The blockage of another artery by this travelling clot is called **embolism**. The coronary arteries are narrower than most others, so emboli tend to become lodged there and cause myocardial infarctions.

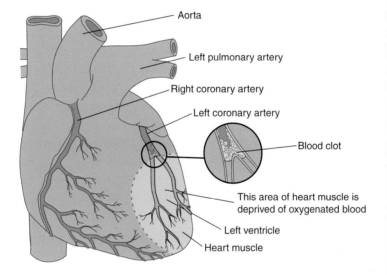

Aorta

Left pulmonary artery

Right coronary artery

Left coronary artery

Blood clot

This area of heart muscle is deprived of oxygenated blood

Left ventricle

Heart muscle

Figure 7.7 A blocked heart artery

Since 'we are what we eat', we should try to avoid food that is high in saturated fat because eating saturated fats increases the rate at which atherosclerosis takes place. Hard fats, such as those in butter, cheeses and in animal tissue contain a much higher proportion of saturated fats than vegetable fats and oils. ◄

A process called **atherosclerosis** increases the tendency of blood clots to form inside arteries. In this process, a mixture of fatty substances, together called **atheroma**, becomes laid down under the endothelial lining of an artery.

Figure 7.8 Atheroma in a blood vessel

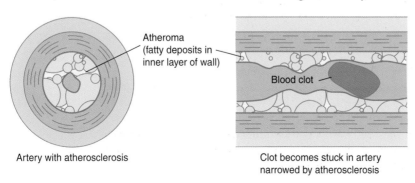

Atheroma (fatty deposits in inner layer of wall)

Blood clot

Artery with atherosclerosis

Clot becomes stuck in artery narrowed by atherosclerosis

The main substance in atheroma is cholesterol. The atheroma subsequently calcifies (absorbs calcium) and becomes harder, forming an **atherosclerotic plaque**. The plaque affects the artery in three key ways:

- It narrows the artery, restricting blood flow — therefore reducing oxygen delivery — and increasing blood pressure.
- It makes it more likely that an embolus will become lodged in the artery.
- It promotes the intrinsic pathway of blood clotting as platelets brush against it.

The extrinsic pathway of clotting occurs when we cut ourselves. The clot formed blocks the wound (see page 137).

Figure 7.9 Causes of clotting in artery

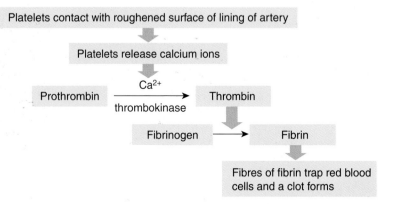

The plaque may also weaken the artery wall, forming an **aneurysm**.

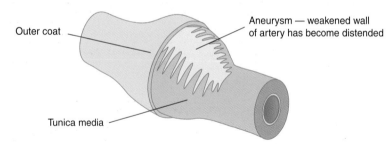

Figure 7.10 An aneurysm

If the aneurysm bursts, serious complications can arise:

- Blood loss from a small aneurysm in an artery in the brain can put pressure on the brain and cause symptoms similar to those of a stroke.
- Blood loss from a large aneurysm in the aorta is nearly always fatal, as too much blood is lost too quickly.

A number of factors influence the incidence of coronary heart disease:

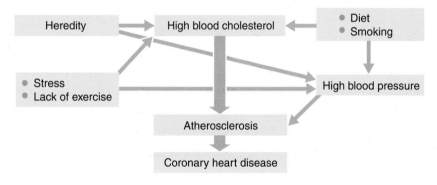

Figure 7.11 Influences on coronary heart disease

One of the most important factors is a genetic disposition towards high cholesterol levels and high blood pressure. You cannot change your genes! However, by modifying your lifestyle, you can reduce the extent to which other factors affect your overall risk.

Box 7.6 Does personality type influence the risk of developing heart disease?

In 1974, two American heart specialists, Friedman and Rosenman, identified two very different personality types. They called these Type A and Type B. People with Type A personality show some or all of the following characteristics:

- competitiveness
- aggression
- very high levels of activity
- hostility
- impatience
- fast talking and thinking

Figure 7.12 Personality types A and B

Type-B personalities are much more relaxed in their attitudes. Friedman and Rosenman identified 3524 Californian men between the ages of 39 and 59 years who were free from any signs of coronary heart disease. The men were given a questionnaire, which was used to assign them to Type A or Type B personalities.

The participants were then monitored for 8.5 years. At the end of this time, Friedman and Rosenman found that 230 of the men (7%) had developed signs of coronary heart disease. Of these, 70% had a Type A personality.

Many of the characteristics of a Type A personality have been linked with stress. It is also known that stress produces a bodily response that increases the levels of plasma lipids (including cholesterol) and of the hormones adrenaline and noradrenaline. These hormones increase heart rate and blood pressure, which are two of the predisposing factors to coronary heart disease (see Figure 7.11).

Type A quiz (check your personality)		
Answer true or false.		
(1) I don't let people know when I am angry.	T	F
(2) Most people are generally selfish and self-centred.	T	F
(3) Other people consider me a relaxed person.	T	F
(4) I feel anxious whenever I am idle.	T	F
(5) I can usually trust the people I work with.	T	F
(6) I think people are basically good.	T	F
(7) I become irritated when I must wait for something.	T	F
(8) I stay calm in emergency situations.	T	F
(9) I am usually patient while waiting for someone.	T	F
(10) I tend to keep my feelings to myself.	T	F
(11) I enjoy working against deadlines.	T	F
(12) It is important to take vacations regularly.	T	F
(13) I tend to concentrate on one problem at a time.	T	F
(14) I really can't trust other people.	T	F
(15) Other people have mentioned my hostility.	T	F
(16) I listen to the ideas of other people.	T	F
(17) I often race against time even when there is no reason to.	T	F
(18) I often feel suspicious toward others.	T	F
(19) I do not work well under deadlines.	T	F
(20) I try to relax when things slow down.	T	F

Figure 7.13 Friedman and Rosenman questionnaire

This study produces findings that are correlational. due to lack of proper control of variables, cause and effect have not been definitely identified. Although a relatively large number of men were involved, it is still only a small fraction of the population. The men were all middle-aged (39–59), which is an age group for which the risk of coronary heart disease in the general population increases significantly anyway.

Other studies have produced conflicting results, so the picture is not clear. Some researchers recognised that there was a lack of precision in the original study and focused on just one or two aspects of the Type A personality in their research. For example, in 1986 Matthews and Haynes investigated hostility and the suppression of anger. This produced a much higher correlation with coronary heart disease than the original study by Friedman and Rosenman. Nevertheless, the original study 'pointed the way' for other scientists to follow.

What personality type are you (see quiz)?

Score 5 points for the following answers:
- True: 1, 2, 4, 7, 10, 11, 14, 15, 17 and 18
- False: 3, 5, 6, 8, 9, 12, 13, 16, 19 and 20

0–15	definite Type B
20–35	Type B
40–60	both Type A and Type B traits
65–80	Type A
85–100	definite Type A

From Friedman, M. and Rosenman, R. H. (1974)
Type A Behaviour and Your Heart, Knopf.

The biological basis of cancer

Cancers are formed when a cell divides in an uncontrolled fashion. The cells derived from this form a **clone** of genetically identical cells that also divide in an uncontrolled way. Very soon, a mass of cells called a **tumour** is formed. Most tumours are **benign**. These are usually harmless, although they may cause problems because of *where* they grow. **Malignant** tumours divide in a more uncontrolled way and are much more dangerous. It is these tumours that we call **cancers**.

Benign and malignant tumours

Benign tumours and malignant tumours differ in a number of important ways:
- Benign tumours usually grow much more slowly than malignant tumours.
- Benign tumours usually remain encased within a fibrous capsule and do not invade the tissue in which they originated. The boundaries of malignant tumours are much less defined and the cells frequently invade the tissue in which they originate.
- Benign tumours rarely show metastasis, i.e. they do not usually spread to other parts of the body. Many malignant tumours do metastasise and cause secondary cancers. These secondary cancers are frequently less well defined than the primary cancers, as the cancerous cells enter the 'secondary organ' in a number of different places.
- Malignant tumours often stimulate the development of a blood supply to the tumour.

For these reasons, it is often possible to remove benign tumours successfully by surgery, whereas complete removal of a malignant tumour is much more difficult and success is much less certain. However, even benign tumours can be dangerous. A benign tumour in the brain can exert pressure and affect brain function.

For example, pressure on the parts associated with the eyes can cause loss of vision. Pressure on certain motor areas can cause muscle weakness. Benign tumours may also exert pressure on the blood supply to an organ, restricting the blood flow.

Box 7.7 Smoking and lung cancer

In 1951, Dr. Richard Doll led a team that began a study of the effects of smoking in British doctors. The study was a **prospective study**, because it declared in advance its aims and asked doctors to keep records of their smoking habits and general health. The study is still ongoing and has produced some of the most convincing epidemiological evidence linking smoking to lung cancer, other respiratory diseases and coronary heart disease. Table 7.3 shows the incidence of lung cancer, cancers of other parts of the gas exchange system and chronic obstructive pulmonary disease in British doctors, as reported in 2001.

Chronic obstructive pulmonary disorder' is the collective term given to conditions such as chronic bronchitis, emphysema and some other disorders that result in an impairment of air flow into and ◀ out of the lungs.

Cause of death	Lifelong non-smoker	Former smoker (mean)	Current smoker (mean)	Current – cigarettes per day 1–14	Current – cigarettes per day 15–24	Current – cigarettes per day ≥ 25
Cancer of lung	0.17	0.68	2.49	1.31	2.33	4.17
Cancer of mouth, throat, oesophagus and larynx	0.09	0.26	0.6	0.36	0.47	1.06
Chronic obstructive pulmonary disorder	0.11	0.64	1.56	1.04	1.41	2.61
Other respiratory disease	1.27	1.70	2.39	1.76	2.65	3.11

Other data in the report show significantly reduced life expectancies for smokers, compared with non-smokers.

There is a wealth of other epidemiological evidence linking smoking to lung cancer. The graphs below compare the smoking habits of UK men in the twentieth century with the incidence of lung cancer.

Table 7.3 Deaths per 1000 doctors per year

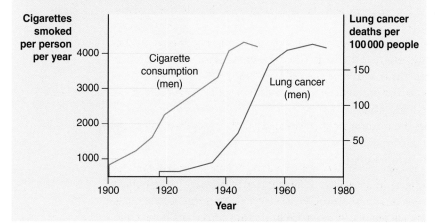

Figure 7.14 Graphs comparing the smoking habits of UK men in the twentieth century with the incidence of lung cancer

The parallels between the smoking and cancer graphs cannot be just coincidence — or can it? Cigarette manufacturers certainly argued for years that it was. Yet again, the data is correlational.

Even carefully controlled experiments on animals failed to convince the cigarette manufacturers. You must decide for yourself whether it was ethically acceptable to subject different species of mammals to cigarette smoke at a range of concentrations, but, for most people, these experiments established the link between concentration of cigarette smoke inhaled and incidence of lung cancer. However, the cigarette manufacturers argued that the data were not acceptable because they came from several non-human species, and the concentrations were often much higher than those experienced by human smokers.

Later, analysis of cigarette smoke showed that it contains substances called benzopyrenes.

Benzopyrene has been shown to bind with DNA in a tumour suppressor gene known as P53. It has also been shown that it causes mutations at several points ('hot-spots') within this gene. These mutations are exactly the ones that are found in most lung cancers. At last, definitive experimental evidence showed that chemicals in cigarette smoke cause lung cancer.

Unlike infectious diseases, improvements in living conditions and diet have little effect on the incidence and survival rates from lung cancer. It is nearly always fatal.

Barbecued food and burnt toast also contain higher than normal levels of benzopyrenes; the risk posed by these foods is not clear.

(a)

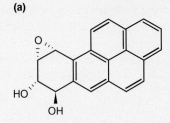

(b)

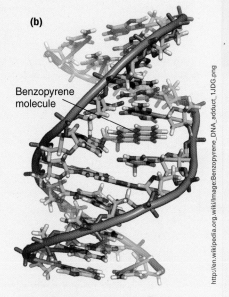

Benzopyrene molecule

http://en.wikipedia.org.wiki/Image:Benzopyrene_DNA_adduct_1JDG.png

Figure 7.15 (a) A benzopyrene molecule (b) The benzopyrene molecule binds to DNA and distorts the double helix

The biological basis of emphysema

Emphysema is one form of chronic obstructive pulmonary disorder (COPD). In all forms of COPD, there is an impairment of flow of air into and out of the lungs.

One form of emphysema is inherited, but most cases occur as a result of smoking. Cigarette smoke passes down the bronchi and bronchioles into the alveoli. Here, the mixture of toxins provokes an immune response called an **inflammatory response**, which ultimately leads to the breakdown of the walls of the alveoli.

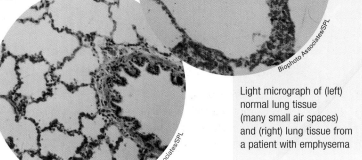

Light micrograph of (left) normal lung tissue (many small air spaces) and (right) lung tissue from a patient with emphysema

Phagocytic white blood cells release the enzyme **elastase** that breaks down a protein called **elastin**, which is found in connective tissue between the airways in the lungs. The purpose of the elastin is to help exhale air. As its name suggests, elastin molecules are 'elastic'. When we inhale, the airways and alveoli dilate and the elastin is stretched. When we exhale, the elastin molecules 'recoil' and help to force air out of the alveoli. So why do phagocytes destroy elastin?

Phagocytes destroy elastin so that they can move more easily through lung tissue in order to engulf and kill microorganisms, should an infection occur. Normally, excessive destruction of elastin is prevented because cells in the lungs secrete a substance called A1AT that inhibits the action of elastase. The digested elastin is soon replaced.

Cigarette smoke contains several oxidants that stop the production of A1AT. Therefore, as phagocytes are repeatedly drawn to the area, the secretion of elastase continues. Over a long period of time, this has two main effects:

- The elastic tissue is damaged, so the elasticity of the tissue around the airways, including the alveoli, is reduced. This makes it harder for a sufferer to exhale effectively.
- Elastase is a protein-digesting enzyme, so it causes damage to the walls of the alveoli. Over time, the walls of the alveoli are broken down so that several alveoli merge into one larger cavity. This reduces the surface area available for gas exchange.

(a) 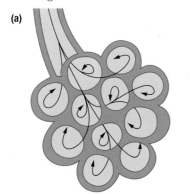 **(b)**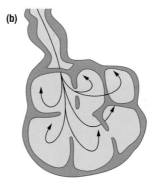

Figure 7.16 (a) Healthy alveoli (b) Alveoli from a person with emphysema

In the early stages of the disease, a sufferer hyperventilates slightly to compensate for the reduced airflow and reduced gas exchange. However, as time passes and the extent of the damage increases, this becomes inadequate and the sufferer becomes more and more disabled. In cases of advanced emphysema, getting out of bed in the morning requires a major effort. Emphysema kills over 20 000 people each year in the UK.

The biological basis of pulmonary fibrosis

Pulmonary fibrosis is a condition in which the tissue between the bronchioles and alveoli becomes scarred as a result of some substance entering the lung. In many cases, it is not possible to identify the precise cause of the fibrosis — there are over 140 different causes listed, but the effect is the same. The development of scar tissue in the lungs twists the alveoli and bronchi out of shape, compressing some and stretching others, and reduces the elasticity of the tissue between the airways. As with emphysema, the symptom of breathlessness is due to poor airflow and gas exchange, although the cause is different.

Pulmonary fibrosis is often linked with occupation. Any job that involves producing small solid particles which might be inhaled will increase the risk of fibrosis, for example:

● jobs that involve grinding stone or metal
● working with asbestos — some kinds of asbestos release small, sharp fibres
● working with mouldy hay — the hyphae of some fungi trigger an inflammatory response that brings about fibrosis; in this case it is known as 'farmer's lung'

The biological basis of asthma

Asthma is a condition in which the bronchioles become chronically inflamed. The exact cause of the initial inflammation is not clear, but it is thought that the following might be causes:

● A genetic risk — there is no known 'asthma gene' but a predisposition towards asthma seems to be more common in some families than in others.
● Exposure to certain allergens at an early age — for example, tobacco smoke, pollen grains, dust mites and other particles that can be inhaled. In some people, these produce an allergic response.

◀ One hypothesis of the cause of asthma is that people in developed countries (where asthma is much more common) are no longer exposed to many of the diseases that their developing immune systems would once have encountered. This makes their immune systems over-react to some harmless foreign particles, causing the initial inflammation.

White blood cells called mast cells invade the bronchioles and release histamines. This causes inflammation. The inflammation of the bronchioles makes them much more sensitive to the triggers that can provoke an asthma attack. In an asthma attack:

● The smooth muscle around the bronchioles contracts, making the lumen narrower.
● Cells lining the bronchioles secrete more mucus than normal, further obstructing the flow of air.
● The restricted air flow makes it more difficult to breathe; the person starts to breathe more quickly but each breath is much shallower.
● As a result of the restricted airflow, gas exchange in the alveoli is reduced.

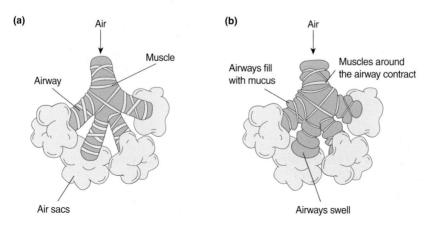

Figure 7.17 External views of airways (a) Before an asthma attack (b) During an asthma attack

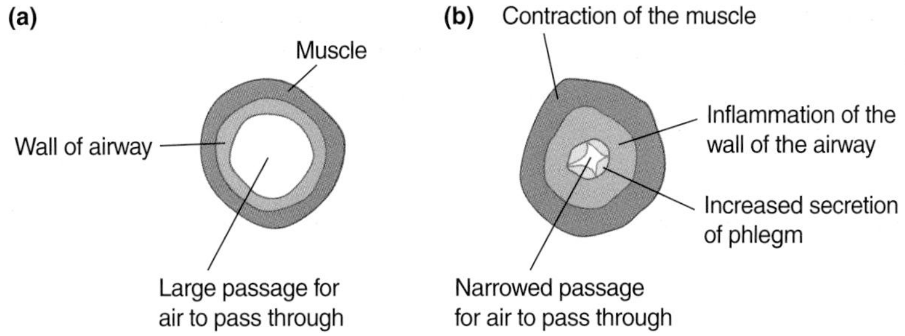

(a)

Muscle

Wall of airway

Large passage for
air to pass through

(b) Contraction of the muscle

Inflammation of the
wall of the airway

Increased secretion
of phlegm

Narrowed passage
for air to pass through

Figure 7.18 Cross-sections
of airways (a) before an
asthma attack and
(b) during an asthma attack

The triggers that can cause an attack include:
- some diseases, particularly the common cold and 'flu
- exposure to fumes, smoke and dust
- exposure to allergens, such as pollen, animal fur, some medicines and some foods
- exercise – especially in cold dry air
- emotions (laughing or crying hard) and stress

Asthma cannot be cured and although, in some cases, it 'resolves' as the person develops, this is not always the case. It can, however, be treated, most commonly by the use of inhalers.

There are two main types of inhaler:
- **Relievers** are used during an attack. They release substances that relieve the symptoms. They cause the smooth muscle to relax so that the bronchioles dilate (become wider). Breathing returns to normal.
- **Preventers** are used daily, whether or not an attack occurs. They release substances (often including steroid drugs) that reduce the under-lying inflammation of the bronchioles. After a period of time (usually about six weeks), the inflammation is sufficiently reduced to make an asthma attack much less likely.

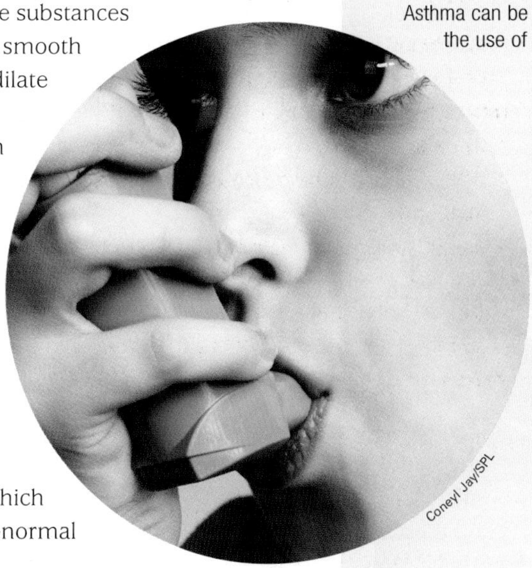

In the UK, 5.2 million people are currently being treated for asthma, 1.1 million of whom are children

Asthma can be treated by the use of inhalers

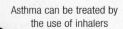

Coneyl Jay/SPL

Summary

Disease is a condition with a specific cause in which part or all of the body is made to function in a non-normal and less efficient manner

Disease

Disease can be caused by:
- pathogenic microorganisms, which cause infectious diseases (e.g. cholera and TB) that may also be communicable (e.g. TB)
- our genes (genetic/inherited diseases)
- our lifestyle and working conditions (human-induced diseases, e.g. fibrosis)

- degenerative processes that occur as we age (degenerative diseases, e.g. arthritis)
- deficiencies in our diet (deficiency diseases, e.g. scurvy and kwashiorkor)
- social activities (social diseases, e.g. alcoholism and drug addiction)

Infectious disease

- Most infectious diseases affecting humans are caused by bacteria, viruses and fungi.
- Cholera is caused by a bacterium that enters the body in polluted water or, occasionally, in food. Once in the small intestine, the bacterium multiplies and releases toxins that cause:
  - an increase in permeability of the intestinal epithelium to ions which leave the cells and enter the lumen of the intestine,
  - water loss from the cells by osmosis due to the resulting osmotic gradient
  - dehydration due to loss of water by osmosis from the blood
  - diarrhoea as the large intestine moves the contents through too quickly to reabsorb the lost water.
- Cholera can be treated by oral rehydration therapy; this replaces the lost ions and water, but does not cure the disease.
- Pulmonary tuberculosis is caused by a bacterium that is usually spread by droplet infection.
- The bacteria causing TB multiply inside phagocytes in the lungs forming tubercles and breaking down lung tissue.
- Symptoms of TB include:
  - chest pain and coughing up blood
  - a cough that lasts for more than three weeks
  - chills, fever and night sweats
  - loss of appetite and loss of weight
  - fatigue
- TB is treated by a combination of four antibiotics because resistant strains of the bacterium exist that are resistant to two, or sometimes three, antibiotics.
- The BCG vaccine is used to prevent TB; it is up to 80% effective in children

Diseases reflecting the effects of lifestyle

- Coronary heart disease occurs when one of the coronary arteries is blocked, depriving heart muscle of oxygenated blood and eventually causing a myocardial infarction.
- Atherosclerosis is the laying down of fatty substances in the lining of arteries. This narrows the arteries and increases:
  - blood pressure (which further increases the rate at which atherosclerosis takes place)
  - the likelihood of clot formation in the artery; if the clot breaks up it forms an embolism which may travel to a coronary artery and block it
  - the likelihood of an aneurysm developing
- Coronary heart disease is a multifactorial condition in which diet, genetic factors, exercise, stress, smoking, blood cholesterol levels, blood pressure and rate of atherosclerosis all influence how the condition develops.

- A tumour is a clone of cells formed when genes in a cell mutate and the cell begins to divide in an uncontrolled manner. These mutations occur in proto-oncogenes and tumour suppressor genes.
- Benign tumours are slow-growing and are contained within a membrane; they are often harmless, but as they increase in size they may put pressure on other structures.
- Malignant tumours are cancerous; they grow much faster than benign tumours, do not have defined boundaries and may show metastasis (spread to other organs).
- Cigarette smoking causes lung cancer by inducing mutations in the P53 tumour suppressor gene.
- Smoking cigarettes also causes emphysema; the tissues in the airway become inflamed and phagocytes release elastase that digests elastin and other proteins in lung tissue, causing:
  - a loss of elasticity in lung tissue, making it harder to exhale
  - damage to the walls of the alveoli, causing gaps to appear in the lungs and reducing the surface area of the alveoli that is available for gas exchange
- Pulmonary fibrosis results from inhaling tiny solid particles that damage the lung tissue and lead to the formation of fibrous, non-elastic scar tissue between the airways. The fibrous tissue distorts the alveoli and bronchi and makes the lung tissue less elastic; gas exchange and breathing are both more difficult.
- Asthma is a condition that results from a chronic inflammation of the bronchi and bronchioles, which makes them sensitive to a number of asthma 'triggers'.
- In an asthma attack, the triggers (e.g. pollen, dust mites, stress, emotion) cause the smooth muscle around the bronchioles to contract, narrowing the airways. Extra mucus is also produced, narrowing the airways further. Breathing becomes difficult due to the reduced diameter of the bronchioles.
- Asthma is treated with reliever inhalers (that relieve the symptoms of an attack by causing the bronchioles to dilate) and preventer inhalers (that reduce the inflammation and so reduce the effect of the triggers).

Questions

Multiple-choice

1 Disease is best described as:
  A an absence of health
  B a specific condition caused by an infectious microorganism
  C a specific condition in which part or all of the body functions in a non-normal and less efficient manner
  D an absence of health caused by an infectious microorganism

2 Cholera is:
  A infectious and communicable
  B infectious, but not communicable
  C not infectious, but communicable
  D neither infectious nor communicable

3 Evidence concerning the change in patterns of the incidence of disease often suggests a link to a possible cause. Such evidence is said to be:

A observational

B correlational

C pandemic

D experimental

4 Coronary heart disease is:

A a non-infectious disease

B a multifactorial condition

C partly due to factors that are inherited

D all of the above

5 Coronary thrombosis means:

A having a heart attack

B the narrowing of a coronary artery by atherosclerosis

C the formation of a blood clot in a coronary artery

D all of the above

6 In an asthma attack:

A bronchioles dilate and mucus secretion increases

B bronchioles constrict and mucus secretion increases

C bronchioles constrict and mucus secretion decreases

D bronchioles dilate and mucus secretion decreases

7 In pulmonary fibrosis:

A elastic tissue is digested and scar tissue is formed

B elastic tissue is digested and scar tissue is digested

C alveoli are distorted and scar tissue is digested

D alveoli are distorted and scar tissue is formed

8 The spread of pulmonary tuberculosis:

A occurs by droplet infection

B has been reduced by improved diet and living conditions

C is increased among people with AIDS

D all of the above

9 Benign and malignant tumours differ in that:

A benign tumours are slower growing and are enclosed in a membrane

B benign tumours are slower growing and do not have defined boundaries

C benign tumours are quicker growing and do not have defined boundaries

D benign tumours are quicker growing and are enclosed in a membrane

10 In emphysema:

A the total surface area of the alveoli is increased and the lung tissue is more elastic

B the total surface area of the alveoli is increased and the lung tissue is less elastic

C the total surface area of the alveoli is decreased and the lung tissue is less elastic

D the total surface area of the alveoli is decreased and the lung tissue is more elastic

Examination-style

1 An asthma attack occurs when an asthma trigger affects the inflamed bronchioles.

  (a) Name two asthma triggers.   *(2 marks)*

  (b) Describe two changes that occur in an asthma attack and explain how these changes affect lung function.   *(4 marks)*

  (c) The substances contained in reliever inhalers are called 'bronchodilators'. Explain why they are given this name.   *(2 marks)*

Total: 8 marks

2 In 1951, Dr Richard Doll began a survey into the health of British doctors and their smoking habits. The graphs show the survival rates from age 35 of doctors born at different times.

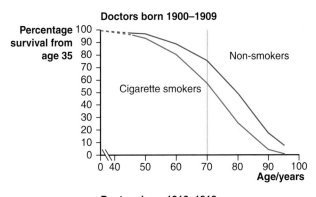

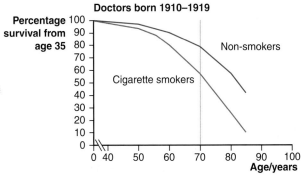

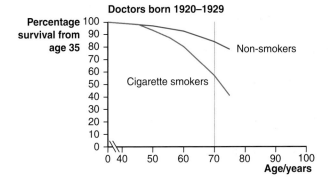

(a) (i) Describe the change in survival rates to age 70 of
non-smokers born at different times. *(2 marks)*

(ii) Suggest a reason for the differences in survival rates
that you described in your answer to (i). *(2 marks)*

(b) The survival rates to age 70 for the three groups of doctors
who smoked is the same. Suggest an explanation for this. *(3 marks)*

(c) Explain how a cancer develops. *(5 marks)*

Total: 12 marks

3 The graph shows the prevalence of cigarette smoking in the UK by different
age groups from 1974 to 2005.

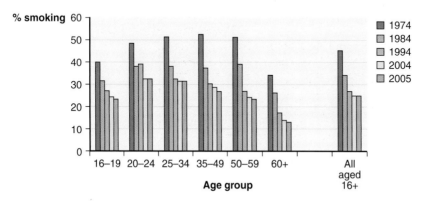

(a) (i) Describe the change in prevalence of smoking between
1974 and 2005. *(1 mark)*

(ii) Which age group showed the biggest change between
1974 and 1984? *(1 mark)*

(b) Calculate the mean percentage change in the prevalence
of cigarette smoking between 1974 and 2005. *(2 marks)*

(c) Name three diseases that are known to be linked to
smoking cigarettes. *(3 marks)*

Total: 7 marks

4 The diagram shows an external view of the heart.

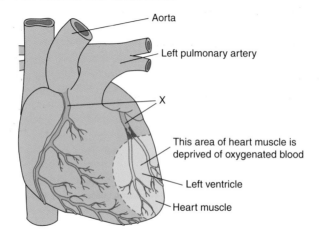

(a) (i) Name the structures labelled X. *(1 mark)*

(ii) What is the function of the structures labelled X? *(3 marks)*

(iii) Explain how a blood clot formed elsewhere in the body may result in the blockage of one of these structures. *(3 marks)*

(b) List three factors that increase the risk of developing coronary heart disease. For each factor you name, explain how it brings about the increased risk. *(3 marks)*

Total: 10 marks

5 The table shows, in order, the ten leading causes of death at any age in Los Angeles county in the USA in 2003. It also shows their ranking as the cause of premature death (before the age of 55).

Cause of death	Number of deaths	Premature (before 55 years) death ranking
Total deaths, all causes	61 026	—
Coronary heart disease	16 320	1
Stroke	4249	8
Lung cancer	3150	5
Emphysema	2796	13
Pneumonia and influenza	2419	18
Diabetes	2178	9
Colon cancer	1450	12
Alzheimer's disease	1285	51
Breast cancer	1084	11
Murder	1066	2

(a) Calculate the percentage of deaths due to coronary heart disease. *(2 marks)*

(b) (i) How many infectious diseases are listed in the top ten causes of death? *(1 mark)*

(ii) Suggest why there is only a small number of infectious diseases in the list. *(4 marks)*

(c) Suggest explanations for the difference in the two ranks for:

(i) emphysema *(3 marks)*

(ii) pneumonia and influenza *(2 marks)*

(d) Suggest why pulmonary tuberculosis does not appear in this list. *(3 marks)*

Total: 15 marks

Chapter 8

How do we resist infection?

This chapter covers:
- the different levels of defence against infection in the human body
- the nature of antigens and antibodies
- the antibody-mediated immune response
- the cell-mediated immune response
- primary and secondary immune responses
- vaccination
- the importance of monoclonal antibodies

Most of us suffer from diseases at some time or other. Sometimes we just feel slightly unwell for a few days; on other occasions we may have to take time off school or work and possibly spend a few days in bed. But we generally recover and then, usually, we don't suffer from that disease again — we are immune. So how did we get better? And why are we then usually immune? In this chapter we look at how our bodies are equipped to keep most microorganisms out, and how they respond to, and usually kill, those microorganisms that do manage to enter.

How are we able to resist infection by pathogenic microorganisms?

It is possible to resist infections because of:
- species resistance
- physical and chemical barriers that help to exclude microorganisms from the body
- immune responses that destroy microorganisms that have succeeded in entering the body

What is species resistance?

Species resistance in the case of humans is resistance to disease simply because we are *Homo sapiens*. We do not contract many diseases that are common in other species. For example, a human is unlikely to suffer from Dutch elm disease or from canine distemper. This is usually because the conditions in the human body do not provide a suitable environment for the pathogens that cause these diseases. In other cases, it is because the pathogen cannot easily infect human cells or tissues.

How do we exclude pathogenic microorganisms?

Pathogenic microorganisms might gain entry to our body at any point where there is an 'interface' with the environment. The main methods of excluding microorganisms from the body are shown in Figure 8.1.

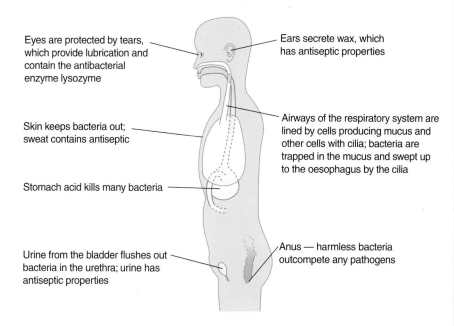

Eyes are protected by tears, which provide lubrication and contain the antibacterial enzyme lysozyme

Ears secrete wax, which has antiseptic properties

Skin keeps bacteria out; sweat contains antiseptic

Airways of the respiratory system are lined by cells producing mucus and other cells with cilia; bacteria are trapped in the mucus and swept up to the oesophagus by the cilia

Stomach acid kills many bacteria

Urine from the bladder flushes out bacteria in the urethra; urine has antiseptic properties

Anus — harmless bacteria outcompete any pathogens

Figure 8.1 Ways of excluding microorganisms

Box 8.1 Blood clotting

Blood clotting at a wound is another important way of keeping out pathogenic microorganisms. The chain of events that leads to blood clotting and the formation of a scab at a wound is called the extrinsic pathway. This is shown in the diagram below.

Damaged tissues release chemicals

Platelets release calcium ions

$$\text{Prothrombin} \xrightarrow[\text{thrombokinase}]{\text{Ca}^{2+}} \text{Thrombin}$$

Fibrinogen $\longrightarrow$ Fibrin

Fibres of fibrin trap red blood cells and a clot forms

When blood clots, red blood cells become trapped in a mesh of fibrin fibres

(× 3000)

Susumu Nishinaga/SPL

How do we destroy pathogenic microorganisms and make ourselves immune to future infections?

If pathogenic microorganisms manage to get past the physical and chemical barriers, the body can put into operation a number of **immune responses**. Some of these are non-specific immune responses; others are specific. Only the specific immune responses give us any lasting immunity.

Non-specific immune responses

Non-specific immune responses are not dependent on the presence of any *particular* foreign antigen in the body, but some foreign antigens must be present to trigger the response.

- **Fever** raises the temperature of the body. This happens because many pathogens release chemicals that stimulate the hypothalamus to 'reset' the body's thermostat to a higher temperature — perhaps to 40°C rather than 37°C. The higher temperature causes more damage to the cells of the pathogen than to the cells of the body. During a fever, because the body thermostat has been reset, attempts to reduce the temperature will be resisted by the body's temperature regulation systems. Once the pathogen has been destroyed, it is usually best to let the fever 'break' naturally.

- **Inflammation** due to infection has four classic signs:
  - redness   – pain
  - swelling   – heat

  The consequence of inflammation is that capillaries in the infected area become more permeable. More white blood cells, antibodies and 'complement proteins' can escape from the capillaries.

- **Phagocytosis** (Figure 8.2) involves the ingestion and subsequent digestion of microorganisms by a range of white blood cells. The most common white blood cells that act as phagocytes are **neutrophils** and **macrophages**. When tissue is damaged by infection, chemicals called mediators are released, which attract phagocytic cells by chemotaxis. Often, they can escape more easily into the tissues because of the increased permeability of the capillaries caused by inflammation. They engulf bacteria by enclosing them in phagosomes formed by pseudopodia. Lysosomes then migrate to the phagosome and secrete hydrolytic enzymes to digest the microorganism.

◄ Complement proteins are a group of about 20 proteins that cause a cascade of reactions leading to the lysis (bursting) of bacterial cells. The complement proteins can also be activated by specific immune responses.

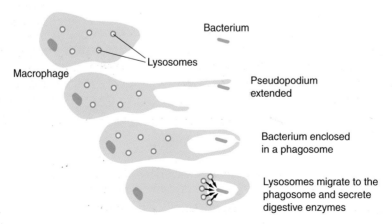

Figure 8.2 Phagocytosis

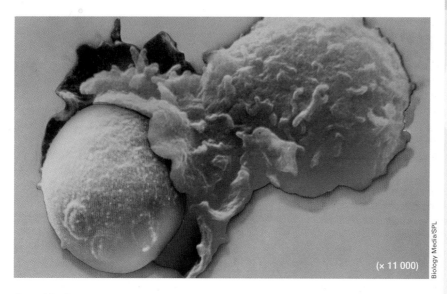

(× 11 000)

Biology Media/SPL

A yeast cell being engulfed by a white blood cell

Specific immune responses

Non-specific responses do not give any lasting immunity to a pathogen. If the same type of bacterium were to enter the body the day after being destroyed by a combination of non-specific responses, the whole process would take place again, in exactly the same way. Specific immune responses, in addition to destroying the pathogen, usually result in an **immunological memory** that gives lasting protection against that particular pathogen. There are two types of specific immune response:

- the humoral, or antibody-mediated, response
- the cell-mediated response

◀ When we have developed this immunological memory, we are said to be immunised against the disease.

Antigens and antibodies

Both the humoral and cell-mediated responses occur when a foreign antigen is detected in the body. Many different types of molecule can act as antigens — proteins, carbohydrates, even DNA — but most antigens are proteins and, very often, glycoproteins.

> An antigen is any molecule that produces an immune response.

The antigen can be the pure chemical itself, or it might be carried on the surface of a pollen grain, a bacterial cell, a virus or a human cell. An immune response is only stimulated when a foreign or **non-self-antigen** is detected in the body. Each person has an individual 'set' of antigens called **self-antigens**. Normally, the cells of the immune system do not attack self-antigens. Often, it is not the whole antigen molecule that stimulates the immune response but certain parts of it, called **antigenic determinants**.

Occasionally, the immune system begins to attack some self-antigens, resulting in autoimmune disease. Iritis (serious inflammation of the iris) can be an autoimmune condition, as can some ◀ types of diabetes.

Production of antibodies in the humoral response

Antibodies belong to a class of proteins called **immunoglobulins**. They are produced by cells derived from B-lymphocytes and each can bind with a specific antigen. Each antibody molecule is basically 'Y'-shaped. Part of the molecule is the same in all antibodies — the constant region. The part that binds with the antigen is different in different antibodies — the variable region (Figure 8.3).

> An antibody is a protein produced by a B-lymphocyte in response to a specific antigen.

The B-lymphocytes that are responsible for the antibody-mediated response develop just before and just after birth, from stem cells in the bone marrow. At this stage they are inactive, but each 'presents' its antibody molecules on the plasma membrane. Each B-lymphocyte produces a slightly different antibody. The inactive B-lymphocytes then migrate to lymph nodes, the liver and spleen where they remain, unless stimulated by an antigen to which their antibody can bind. In this case, the B-lymphocytes become active and divide rapidly by mitosis to form millions of either:

- **plasma cells** — large cells that secrete the antibody into the blood plasma, or
- **memory cells** — B-lymphocytes that are stored in the lymph nodes and, on any subsequent exposure to the same antigen, quickly divide to form many plasma cells

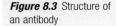

Figure 8.3 Structure of an antibody

◀ Antibodies are able to bind with antigens because the shape of the variable portion of the antibody molecule is complementary to one or more of the antigenic determinants in the antigen.

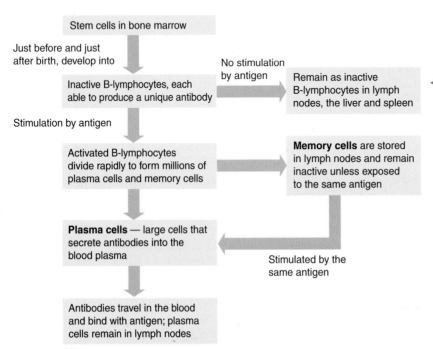

Figure 8.4 Production of antibodies

◀ B-lymphocytes do not have antibodies that bind with self-antigens. When they undergo their initial development, any B-lymphocytes containing antibodies that bind with self-antigens are destroyed. This leaves only those B-lymphocytes that produce antibodies to bind with non-self-antigens.

What triggers a particular B-lymphocyte to develop into plasma cells and memory cells? When the B-lymphocyte encounters a microorganism with an antigen complementary to its antibody, it binds with the antigen and then engulfs it by phagocytosis. Using special proteins, called **major histocompatibility complex (MHC) proteins**, it displays the antigen on its surface. Now another type of lymphocyte, called a **T-helper cell**, which must also have MHC proteins that can bind with the antigen, attaches itself. The helper cell is stimulated to divide and produce chemicals called **cytokines**, which stimulate the B-cell to divide and form plasma cells and memory cells.

The initial production of antibodies by plasma cells derived directly from B-lymphocytes is called the **primary immune response**. Since it depends on the initial stimulation of relatively few cells, it takes some time for the inactive B-lymphocytes to form sufficient plasma cells to make enough antibody to have a real impact on the antigens. During this period, the microorganism carrying the antigen multiplies and illness results.

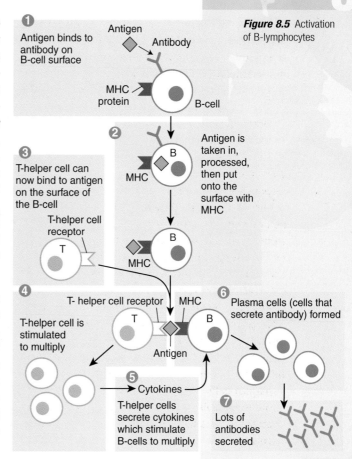

Figure 8.5 Activation of B-lymphocytes

The later production of antibodies by plasma cells derived from memory cells (on reinfection by a microorganism with the same antigen) is the **secondary immune response**. There are many more memory cells than there were original inactive B-lymphocytes, so this response is much faster and many more antibodies are produced.

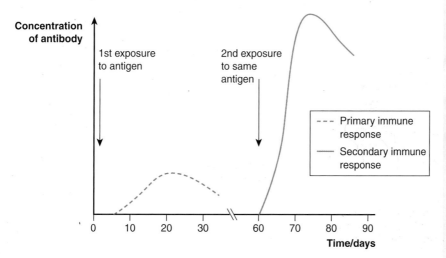

Figure 8.6 Graph comparing primary and secondary immune responses

Action of antibodies

Different antibodies act in different ways. The most common effects on antigens are:

- inactivation — antibodies of this type neutralise toxins produced by bacteria
- agglutination (binding together) — this causes antigen-carrying bacteria to form 'clumps', which are more susceptible to attack by other cells of the immune system (Figure 8.7)
- facilitating phagocytosis
- stimulating the complement proteins, which leads to lysis of the bacterium

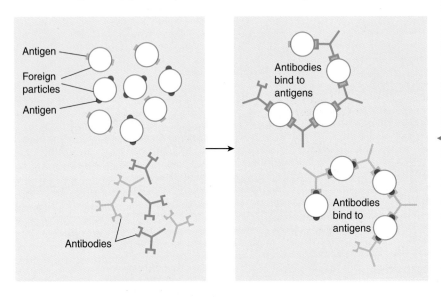

Figure 8.7 Some antibodies bind antigens together, causing clumps of, for example bacteria to form

◀ Immunological research could help to develop our understanding of cancer — why does the body recognise and destroy some cancer cells but not others? In the future, it may also help the development of effective vaccines against malaria and HIV.

The cell-mediated response

The cell-mediated response is brought about by **T-lymphocytes**. These cells are also formed from stem cells in the bone marrow, but they are modified in the thymus gland (hence T-) before migrating to the lymph nodes, liver and spleen. The action of T-lymphocytes is different from that of B-lymphocytes, as is the way in which they are activated. They respond not to microorganisms in the blood, but to body cells that have been invaded by viruses or have become cancerous. Non-self antigens (either from the virus or from the changed identity of the cancer cell) displayed on the plasma membrane stimulate the cell-mediated response.

Two types of T-cells are involved in this response; helper cells are again involved, together with **cytotoxic (or killer) T-cells**. The killer cells bind with the non-self antigens displayed on the affected body cell; the helper cells bind with the same antigens displayed on a macrophage. The helper cell multiplies and the many helper cells that are produced secrete cytokines that cause the killer cells to multiply, forming activated killer cells and memory cells.

The memory cells formed from killer T-cells are not confined to the lymph nodes (as are those formed from B-lymphocytes). T-lymphocyte memory cells circulate freely in the blood and tend to visit most frequently those areas in which they first encountered the antigen. This is known ◀ as **T-cell homing**.

The activated killer cells then destroy the infected cells by either:

- binding to the non-self antigens on the plasma membrane and releasing chemicals that perforate the membrane, killing the cell within seconds, or
- coating the cells with chemicals that mark them out as requiring phagocytosis

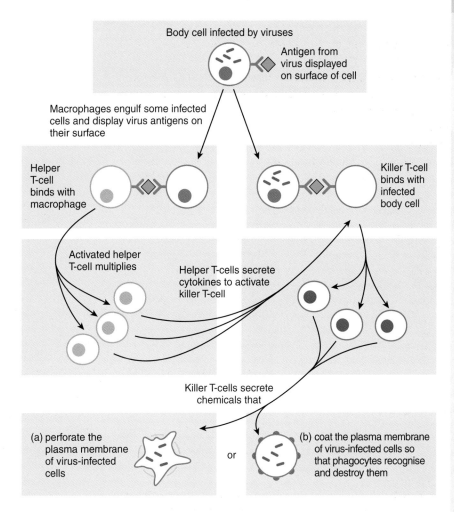

Figure 8.8 Activation of T-lymphocytes

The immunity that results from either the humoral or cell-mediated response to infection is called natural active immunity. However, there are other types of immunity.

Natural immunity arises as a result of natural processes:
- **Natural active immunity** results from immune responses to invading microorganisms.
- **Natural passive immunity** results when antibodies are *acquired*, not made as a result of infection. Antibodies are passed from mother to fetus across the placenta and from mother to baby in the colostrum and milk during breast-feeding.

Artificial immunity is acquired from processes that do not occur naturally:
- **Artificial active immunity** results from **vaccination**. During this process, the body is injected with an 'agent' that carries the same antigens as a particular pathogen. The body produces the same immune response as it would in response to that pathogen.
- **Artificial passive immunity** results from injecting specific antibodies directly.

Artificial active immunity: vaccination

Vaccination involves the injection of a harmless 'agent' that carries the antigens of the microorganism (or its toxin) that causes the disease. Examples include:

- an attenuated (weakened) strain of the microorganism (e.g. vaccines against poliomyelitis, TB and measles)
- dead microorganisms (e.g. vaccines against whooping cough and typhoid fever)
- modified toxins of the causal bacteria (e.g. vaccines against tetanus and diphtheria)
- antigens alone (e.g. some influenza vaccines)
- harmless bacteria or yeasts, genetically engineered to carry the antigens of pathogenic microorganisms (e.g. hepatitis B vaccine)

Using vaccines containing live microorganisms carries a slight risk. In the 1950s, a batch of polio vaccine contained active viruses because it had not been prepared properly. Children who were vaccinated with it contracted polio. Today, risks are better understood and quality-control procedures are more rigorous. The risk is minimal.

Some microorganisms mutate frequently. As a result, the antigens on their surface change, making it almost impossible to develop an effective vaccine. This is true of the virus that causes the common cold. The influenza virus also mutates regularly, but not as frequently as that which causes the common cold. There is usually enough time to develop an effective vaccine before the virus mutates again.

Vaccines have a role to play in preventing the spread of disease. If most of the population is immune to a disease, then there is little chance of the micro-organism finding a suitable host and the disease is maintained at a low level in that population. This is called 'herd immunity'. About 85–90% of people have to be immune to maintain this effect.

Box 8.2 The MMR vaccination

Recently, there has been an increase in the incidence of mumps and measles in the UK. This has happened because of a report in 1998 that the vaccine being used (the MMR triple vaccine against mumps, measles and rubella) was linked with autism. There was a significant decrease in the numbers of young children being vaccinated, with the result that the herd immunity effect was weakened and the incidence of measles and mumps began to rise. In 2006, the incidence of measles was 13 times its 1998 level and that of mumps was 37 times higher. The first death from measles since 1992 occurred in 2005.

Part of the reason for the reaction may have been caused by the way in which the supposed link was reported by the media. A single paper in the *Lancet* (a medical journal) proposed the link with autism. Many governmental reports had found no link, but these were treated with scepticism because of allegations of a 'cover-up'.

As more and more reports have been published showing no link and no new evidence in favour of the link has come to light, confidence in the vaccine is being restored.

What would you need to know to make an informed decision?

What are monoclonal antibodies and why are they important?

When a humoral response occurs, often more than one type of antibody is produced. This is because the microorganism has several different antigens on its surface, each stimulating different B-cells to multiply and produce antibodies. The antibodies in this mixture are called **polyclonal antibodies**. The cells that produce them are clones of each of the original B-cells that was stimulated. Since several cells are stimulated, there are several different clones, each producing different antibodies. This gives the name *polyclonal* (many clones).

By contrast, **monoclonal antibodies** are produced from just one type of cell. Only one type of antibody is produced, so action is specific — the antibodies can target only one particular antigen. This high specificity has attracted the attention of researchers trying to find ways of targeting cancer cells with toxic drugs, in order to kill them without affecting other cells. The principle behind this idea is:
- produce an antibody that will bind with an antigen found only on a cancer cell
- couple a cytotoxic agent (e.g. a radioactive substance) to the antibody
- inject the antibody/cytotoxin complex into the patient and allow it to target the cancer cells and destroy them

Although a strongly radioactive substance is required, the relatively small amounts needed compared with conventional treatments would make the procedure effective, with few side-effects. This does sound like the original 'magic bullet' first proposed by Paul Ehrlich in the early twentieth century. However, there are problems with the procedure. The monoclonal antibody is produced by mice and is, therefore, a mouse (non-human) protein. The human immune system recognises this 'foreign' protein and produces an immune reaction to destroy it.

Box 8.3 How monoclonal antibodies were first produced

In 1951, Henrietta Lacks died from cervical cancer. Cells from her cancer had been saved and cultured. It was discovered that these cancer cells were 'immortal' — provided that the cells were given an appropriate supply of nutrients, they continued to grow and divide. This has since been found to be the case with many cancer cells, including cancerous B-lymphocytes called **myelomas**. These cells will multiply indefinitely, but have lost the ability to synthesise antibodies.

Meanwhile, scientists working in other areas had found ways of making different types of cells fuse. In 1975, Kohler and Milstein found a way of making myeloma cells fuse with mouse spleen cells that had been exposed to a specific antigen. The resultant **hybridoma** cells could produce just one antibody that targeted only the antigen used. Because of the cancerous origin of the hybridoma cell, they too are immortal and can produce the antibody indefinitely. The procedure is summarised in Figure 8.9.

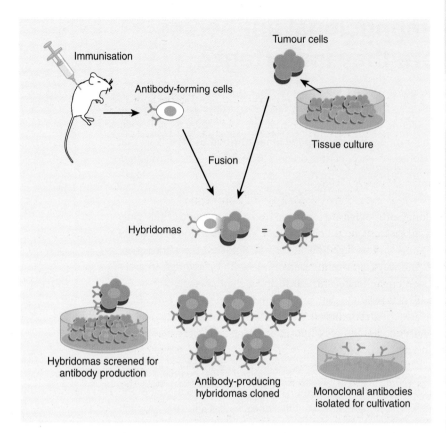

Figure 8.9 Diagram summarising how monoclonal antibodies were first produced

The early monoclonal antibodies did not work as therapeutic agents, so new techniques were developed to 'humanise' the antibodies that were produced, thus reducing the risk of destruction by the body's immune system. Initially, this involved using mouse DNA to code for the variable region of the antibody (the part that would bind with the antigen) and human DNA to code for the non-variable region. These monoclonal antibodies were called chimaeric or humanised antibodies. More recent technologies include the use of genetically engineered mice to produce monoclonal antibodies that have a structure that much more closely resembles that of human antibodies. For example, Herceptin is a monoclonal antibody used to treat breast cancer.

Monoclonal antibodies are now used in a number of treatments, including:
- heart disease
- colon and breast cancer
- leukaemia
- non-Hodgkin lymphoma
- suppressing the immune system in transplant operations

They are also widely used in research for targeting particular antigens. For example, adding a fluorescent or radioactive substance to the monoclonal antibody allows a particular antigen to be detected.

Summary

Immunity

- Resistance to infection by pathogenic microorganisms is possible because of species resistance, physical and chemical barriers that exclude microorganisms from the body, and immune responses that destroy microorganisms that have entered the body.
- Physical and chemical barriers include:
  - the skin (sweat contains an antiseptic)
  - wax in the ears — a physical barrier and also a mild antiseptic
  - mucus in the airways of the lungs
  - tears that lubricate the conjunctiva and contain lysozyme
  - stomach acid that kills many bacteria
  - commensal bacteria in the anus and rectum
  - urine flushing bacteria out of the urethra
  - clotting of blood at a wound
- Non-specific responses to infection include:
  - phagocytosis — microorganisms are engulfed and destroyed by macrophages
  - fever — the increased temperature limits the reproduction of microorganisms
  - inflammation — the increased permeability of capillaries allows more macrophages, antibodies and complement proteins to escape from the capillaries
- In the humoral (antibody-mediated) immune response process:
  - an inactive B-lymphocyte with complementary antibodies on its surface binds to a foreign (non-self) antigen on a microorganism
  - the B-lymphocyte engulfs the antigen and displays it on its surface
  - a helper T-lymphocyte binds with the displayed antigen on the B-lymphocyte and is stimulated to reproduce
  - the T-lymphocytes produce chemicals that stimulate the B-lymphocyte to reproduce and form millions of plasma cells and memory cells
  - the plasma cells secrete antibodies into the bloodstream
  - the memory cells remain in lymph nodes but form plasma cells if the same antigen is detected in the future (due to a second infection by the same microorganism)
- The response to an initial infection is the primary immune response; the response to a subsequent infection is the secondary immune response.
- The secondary immune response is quicker than the primary immune response and produces a higher concentration of antibodies in the blood.
- In the cell-mediated response:
  - cytotoxic (killer) T-cells bind with non-self antigens displayed on an infected body cell, whilst the helper T-cells bind with the same antigens displayed on a macrophage
  - the helper T-cells multiply and secrete cytokines that cause the cytotoxic T-cells to multiply and produce activated cytotoxic cells and memory cells

- the activated cytotoxic cells bind with the non-self antigens on the plasma membrane of the infected cells and either release chemicals to perforate the membrane or release chemicals that mark out the cell to be phagocytosed
- There are several types of immunity:
  - natural active immunity — antibodies and memory cells are formed as a result of infection by pathogenic microorganisms
  - natural passive immunity — antibodies are acquired (e.g. across the placenta and in colostrum/breast milk), not made; no memory cells are formed
  - artificial active immunity — antibodies and memory cells are formed as a result of vaccination
  - artificial passive immunity — antibodies are injected (e.g. in the treatment of tetanus); no memory cells are formed

Vaccination

- Vaccination involves injecting an agent carrying the antigens of a specific disease-causing organism. This may be a weakened (attenuated) strain of the microorganism, dead microorganisms, modified toxins, the pure antigen or genetically modified harmless bacteria or yeasts.

Monoclonal antibodies

- Monoclonal antibodies are formed by just one type of B-cell and are specific to one antigen.
- Their specificity means that monoclonal antibodies can be used to target specific antigens; these could be antigens on cancer cells. The antibodies can be used to carry cytotoxic agents to cancer cells to destroy them without affecting other body cells.

Questions

Multiple-choice

1 Species resistance is:
  A resistance to disease shared with other species
  B a type of specific immune response
  C resistance to diseases that are common in other species
  D a type of non-specific immune response
2 Inflammation allows:
  A more phagocytes to leave the blood
  B fewer complement proteins to leave the blood
  C both of the above
  D neither of the above
3 The cells that are involved in both humoral and cell-mediated immune responses are:
  A B-cells
  B cytotoxic T-cells

C helper T-cells

D virus-infected cells

4 Antigens are, most commonly:

   A proteins or glycoproteins that stimulate an immune response

   B foreign DNA that stimulates an immune response

   C bacteria

   D viruses

5 Antibodies are best described as:

   A J-shaped proteins

   B J-shaped proteins with a common region and a variable region

   C Y-shaped proteins

   D Y-shaped proteins with a common region and a variable region

6 Compared with the secondary antibody-mediated immune response, the primary response is:

   A slower but produces more antibodies

   B quicker and produces more antibodies

   C slower and produces fewer antibodies

   D quicker but produces fewer antibodies

7 Monoclonal antibodies are:

   A specific to just one antigen

   B produced by just one type of cell

   C neither of the above

   D both of the above

8 Plasma cells are:

   A produced from helper T-cells

   B large cells that secrete antibodies

   C retained in the lymph nodes for years

   D capable of secreting chemicals to perforate the plasma membrane of virus-infected cells

9 Hybridoma cells are

   A made by fusing cancer cells with spleen cells from a mouse

   B immortal

   C capable of producing monoclonal antibodies

   D all of the above

10 Phagocytosis always involves

   A the production of antibodies

   B engulfing invading microorganisms

   C both of the above

   D neither of the above

Examination-style

1 (a) When a pathogenic bacterium enters the body, a person is often ill for a few days before starting to feel better. During this time, the person may feel feverish. Explain the reason for:

   (i) the time delay between being infected and starting to
        feel better   *(3 marks)*

   (ii) feeling feverish   *(2 marks)*

(b) AIDS is caused by the human immunodeficiency virus (HIV). HIV infects helper T-lymphocytes. Explain why AIDS sufferers commonly contract other serious illnesses. *(4 marks)*

Total: 9 marks

2 In an investigation into immune responses, some volunteers were injected with an antigen. The level of antibodies against that particular antigen in their bloodstream was monitored over several weeks. Some time later, a second injection of the same antigen was given, together with a second antigen. The levels of antibodies against both antigens were monitored. The results are shown in the graph below.

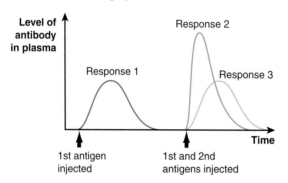

(a) (i) Are responses 1, 2 and 3 primary immune responses or secondary immune responses? *(3 marks)*

   (ii) Explain the difference between response 1 and response 2. *(3 marks)*

(b) Suggest a reason for injecting the second antigen at the same time as the second injection of the first antigen. *(2 marks)*

Total: 8 marks

3 The MMR vaccine offers protection against measles, mumps and rubella. Recently, there has been some concern that the vaccine might have side-effects.

(a) Explain how a vaccine provides protection against disease-causing microorganisms. *(5 marks)*

(b) The table below shows the incidence of some possible complications of the vaccine, compared with the incidence of the same complications following the natural disease.

Complication	Risk per 1 000 000 cases after natural disease	Risk per 1 000 000 cases after MMR
Fits (convulsions)	5000	1000
Meningitis	1000	1
Conditions affecting blood clotting	330	40
Severe allergic responses	–	10
Deaths	100	–

Use the information in the table to evaluate the risks of using the MMR vaccine. *(5 marks)*

Total: 10 marks

4 A number of monoclonal antibodies have been developed to help with the treatment of heart disease. One of these targets a gene that is involved in the metabolism of lipids. The antibody carries a substance that 'knocks out' the gene. Experiments in mice have shown that use of the drug significantly lowers levels of cholesterol in the plasma.

(a) Suggest how monoclonal antibodies are able to target
a particular gene precisely. *(3 marks)*

(b) Suggest how use of the antibody described may reduce
the risk of heart disease. *(3 marks)*

(c) Suggest why the results of this research should be treated
with some caution. *(2 marks)*

(d) Comment on the ethics of this research. *(4 marks)*

Total: 12 marks

Chapter 9

What is the nature of variation?

This chapter covers:
- the differences between interspecific and intraspecific variation
- genetic and environmental variation
- how we investigate variation
- how we record and represent variation
- how we can describe variation mathematically

There is a huge range of living things:

There is a wide variety of animals:

Differences between mammals can be striking:

Primates aren't all the same:

Humans can look different: Even identical twins are not quite the same:

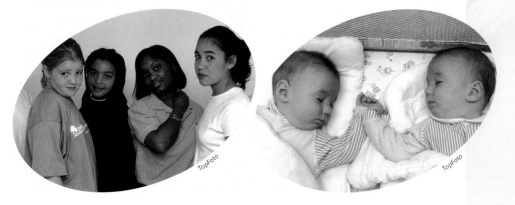

Why are some groups more alike than others?

There are fewer differences between different humans than between humans and other primates. This may seem obvious — but why is it so? What causes the variation?

Humans are all members of the same species — *Homo sapiens* — whereas the different primates are members of different species. The variation within a species is called **intraspecific variation**; that between different species is called **interspecific variation**.

> A species is a group of similar organisms that can interbreed to produce offspring that are fertile.

Some groups share more genes than other groups and so there is less variation between them. Members of a species are usually similar in most ways. This is because they have nearly all their genes in common.

However, humans do not possess the genes that some monkeys have for a prehensile tail. Mammals do not have the genes that reptiles have for the production of scales. Animals do not have the genes that plants have for the production of roots and leaves. As the groups get bigger, the number of genes in common decreases and the variation between the groups increases.

◀ We have 98% of our DNA in common with chimpanzees and 50% in common with bananas!

Genes or environment?

Not all variation is **genetic variation**; factors in the environment cause **environmental variation**. Often the two interact and it can be difficult to determine how much of the observed variation is environmental and how much is genetic.

Box 9.1 Environmental variation in hydrangeas

Hydrangea plants all contain the same pigment in their petals, which is determined by the same gene. However, in some conditions this results in blue flowers and in other conditions in pink or red flowers.

This difference is caused by the environment. The pigment is blue when the plant can absorb enough aluminium ions from the soil, which is only possible if the soil is acidic enough. This is illustrated opposite.

This is a clear example of environment modifying the effect of genes. The gene always codes for the production of the same pigment, but environmental factors modify the nature of the pigment.

Reimar Gaertner/Alamy

The pH of the soil and the availability of aluminium ions influence the colour of the pigment in hydrangea flowers

Box 9.2 Twin studies

'Identical' twins have identical genes. Fraternal or non-identical twins are like any brother and/or sister — they have some genes that are the same and some that are different. By comparing the concordance of certain conditions in different sets of twins, clues can be gained as to the likely influence of genes and the environment.

◀ Concordance is the frequency with which the condition occurs in both twins.

If there is a genetic cause, then concordance ought to be high in identical twins because the same gene will be present in each twin. The concordance should be lower in fraternal twins because they do not necessarily both have the causative gene. If the cause is solely genetic, the concordance will be 1.0 (or 100%), as it is for blood groups. In an investigation into some psychological conditions, researchers measured the concordance of the condition in identical twins and in fraternal twins. The results are shown in Figure 9.1.

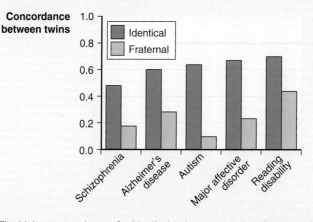

Figure 9.1
The concordance of some conditions in identical and fraternal twins

The higher concordances for identical twins suggest that there may be a genetic component. The fact that this is never 1.0 (100%) suggests that it is not solely genetic. Note that although the results are correlational, they do not *prove* cause and effect.

Height in pea plants is determined by a single gene. There are two versions of this gene; one results in tall plants, the other results in dwarf plants. However, factors in the environment affect just how tall the tall pea plants grow and how tall the dwarf pea plants grow. Once again, genes and environment interact.

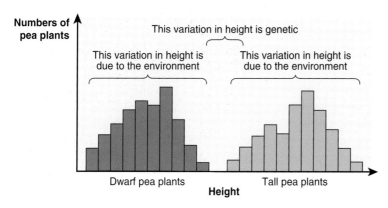

Figure 9.2 Genetic and environmental variation in height in pea plants

How do we investigate variation?

It would be impossible to make a record of all the individuals of a species, since most are spread across wide geographical areas. There are, however, small groups of the same species living in more localised areas. These groups are called **populations**. It is often practicable to estimate the variation within a population.

A population is all the individuals of a species living in the same area at the same time.

Even so, there may be many hundreds, thousands or even millions of individuals in a population and it is clearly not possible to record data about all of them. We have to take a **sample**.

A sample is a subgroup selected to be representative of the group as a whole.

It is also difficult to try to estimate the difference in several features at once. It is simpler to concentrate on just one, or a few, characteristics. So, having decided on which feature(s) to investigate, how do we take our sample? What sort of sample should it be?

As noted earlier, the sample should, as far as possible, be representative of the whole population. How can this be achieved? If we *choose* individuals to be representative of the population, we introduce personal **bias**. By deciding what is representative and what is not, we are prejudging the outcome of the investigation. We need to take a **random sample** of individuals.

To take a random sample of 50 students in a school of 750 would be relatively easy. One way would be to assign each student a number from 1 to 750 and obtain a sample by programming a scientific calculator to select 50 numbers between 1 and 750.

One way to take a random sample of organisms in their natural habitat is to divide the area into equal sections and number them. These could be sites in a field where you are going to lay quadrats or places in a pond where you will obtain a sample of water.

You can then use the random number generator function on a scientific calculator to choose the sites. You can then further 'randomise' exactly how you will select individuals from within these areas.

However, random samples are just that — random. If you and a friend were to each select a random sample of 50 pupils from your school using the method described above, you would almost certainly obtain different samples. The estimate of variation in the feature you are investigating would, therefore, be slightly different. Chance differences in samples are unavoidable and you need to be aware that a sample, chosen at random, may still not be representative of the population.

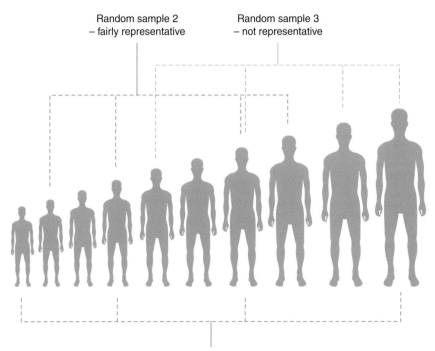

Random sample 2
– fairly representative

Random sample 3
– not representative

Random sample 1 – representative
of the whole population

Figure 9.3 Random samples
are random!

The larger the sample, the more likely it is to represent the population as a whole. However, it then becomes more time-consuming and a balance has to be struck between reliability and time taken.

How do we record and represent variation?

First, you should be aware that there are two types of variation:
- continuous variation
- discontinuous or categoric variation

In continuous variation, a whole range of values or categories is possible and these can be measured — For example, height and body mass in humans, mass of coconut fruits, width of nettle leaves:

In discontinuous variation, only a limited number of discrete (separate) categories is possible. For example, your earlobes are either attached to your cheeks or not, your blood group is A, B, AB or O, a pea plant is either tall or dwarf.

Results from investigations into variation in populations can be recorded in tables and then represented graphically. For example, suppose we investigated human height and earlobe attachment in a sample of 17-year-old students. The results would be recorded in a table — for example, Table 9.1.

Student	Height/cm	Earlobe attached: yes/no
A	157	Y
B	176	Y
C	144	N
D	163	Y
E	165	N
F	159	Y

Table 9.1 Recording the heights and earlobe attachment of students

The results for earlobe attachment could be represented in a **bar chart**:

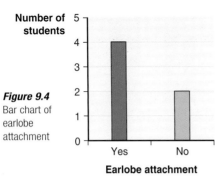

Figure 9.4 Bar chart of earlobe attachment

With height, the situation is rather more complex. Taking the sample of 50 students, it would look confusing if we were to plot all 50 heights on one graph. To get around this, the range of heights is divided into class intervals and the number of students that fall into each interval is noted on a tally chart (Table 9.2).

Table 9.2 A tally chart of the heights of 50 students

Class interval /cm	Tally	Number of students in interval
145–149	/	1
150–154	//	2
155–159	////	4
160–164	₊₊₊ ///	8
165–169	₊₊₊ ₊₊₊ ////	14
170–174	₊₊₊ ₊₊₊ /	11
175–179	₊₊₊ /	6
180–185	///	3
185–189	/	1

Now, we can produce a **histogram** of the results in Table 9.2.

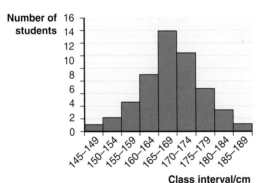

Figure 9.5 Histogram of height in 17-year-old students

Box 9.3 Histograms and bar charts

In the bar chart (Figure 9.4), the bars are separate to show that the different types are not part of a continuous range. In the histogram (Figure 9.5), the bars touch. This is to show that all the values form part of a continuous range. We use bar charts to represent discontinuous (categoric) variation and histograms to represent continuous variation.

A distribution curve can be obtained from a histogram. If a larger sample had been used, the graph would be more symmetrical, as in Figure 9.6.

Figure 9.6 shows a **normal distribution curve**. Features showing continuous variation are often normally distributed.

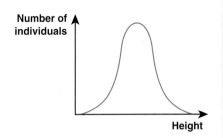

Figure 9.6 A normal distribution curve

How can we describe a normal distribution mathematically?

The diagrams in Figure 9.7 represent distributions of leaf area in two populations of nettle plants.

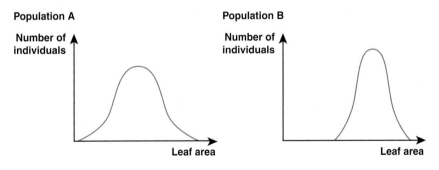

Figure 9.7 Leaf area in two populations of nettle plants

In both areas, the leaf area is normally distributed. However, the curves are very different. To describe a normal distribution we need to answer two questions:
- What is the average or typical value?
- How wide is the spread of values around this typical value?

For the first, we calculate the **arithmetic mean**, $\bar{x}$, of the values: add together the different values in the sample and divide by the number in the sample. The formula for this calculation is:

$$\bar{x} = \frac{\Sigma x}{n}$$

where $\bar{x}$ = the mean value

Σx = the sum of all the individual values

n = the number in the sample

To measure the spread about the mean, we calculate the **standard deviation** (σ). In a perfect normal distribution, 34% of all values lie within the value of the standard deviation above the mean and 34% lie within the value of the standard deviation below the mean. This is shown in Figure 9.8.

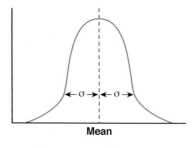

Figure 9.8 The mean and standard deviation of a typical normal distribution

The standard deviation is more useful than just looking at the overall range of values, because this can include one or two freakish values at each extreme. The range is defined by the extremes; it does not matter where the intervening values lie or how many of them there are, the range is unaffected. However, the standard deviation takes into account the extent of the spread of each value from the mean.

To calculate the standard deviation, we must:
- calculate the mean,
- calculate the difference between the mean and each individual value, $(x - \bar{x})$
- square each of these values, $(x - \bar{x})^2$

The standard deviation (σ) is then given by the formula:

$$\sigma = \frac{\Sigma(x - \bar{x})^2}{n - 1}$$

The two normal distributions shown in Figure 9.9 have the same means but their standard deviations are very different.

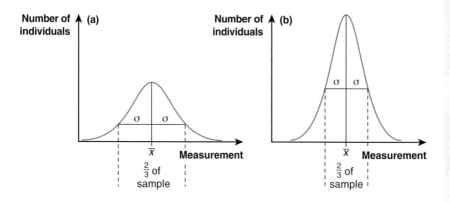

Figure 9.9 Standard deviation

In normal distribution (a), the sample is quite varied. Many individual values are far from the mean, resulting in a large standard deviation.

Normal distribution (b) has the same mean value as (a), but the range of individual values is much smaller. Because there is less variation, the standard deviation is smaller.

The calculation of the standard deviation uses the difference of each value from the mean. If these differences are small, as when there is less variation, the standard deviation is also small.

Therefore, by using the mean and the standard deviation, a normal distribution can be described mathematically. We know the typical value (the mean) and the amount of variability in the sample (the standard deviation).

Summary

- Individuals from the same species have more genes in common and, therefore, show less variation, than individuals from different species. Variation within a species is intraspecific variation; variation between species is interspecific variation.
- Variation can be caused by genes (genetic variation) or by the environment (environmental variation). Frequently, genes and the environment interact.
- To investigate variation in a population, a random sample must be taken to eliminate bias.
- Continuous variation is variation in which there is a continuous range of values — for example human height and body mass. Discontinuous variation is variation in which only a limited number of discrete (separate) categories is possible, such as blood type.
- Continuous variation can be represented by histograms; discontinuous variation by bar charts.
- Continuously variable features are often distributed normally.
- In a normal distribution, the mean is a measure of the typical value. Standard deviation is a measure of how widely the values are spread about the mean; it is a measure of the variability of the sample.

Questions

Multiple-choice

1 Variation between different species is:
   A interspecific
   B continuous
   C intraspecific
   D environmental

2 The total of all the genes of a species is its:
   A genotype
   B gene frequency
   C gene pool
   D gene sequence

3 If a feature is determined solely by the genes, the concordance between identical twins is:
   A 100%
   B 75%
   C 25%
   D 0%

4 When sampling from a population, the sample taken should be random in order to:
   A avoid bias
   B prevent prejudging the outcome
   C both A and B
   D neither A nor B

5 Two random samples of the same population:
   A can never be the same
   B will always be the same
   C will probably be the same
   D are unlikely to be the same

6 Two populations that have the same mean but different standard deviations have:
   A the same typical value and the same variability
   B different typical values and the same variability
   C different typical values and different variability
   D the same typical value and different variability

7 It is preferable to use the standard deviation rather than the range because:
   A it is less affected by extreme values
   B it takes into account all values
   C neither A nor B
   D both A and B

8 To represent discontinuous variation graphically, we would plot a:
   A bar chart
   B histogram
   C line graph
   D scattergram

9 The variation in human height is an example of:
   A interspecific, continuous variation
   B interspecific discontinuous variation
   C intraspecific discontinuous variation
   D intraspecific continuous variation

10 A population is:
   A all the individuals of one species in a certain area
   B all the individuals of all species in a certain area
   C a sample of one species in a certain area
   D a sample of all species in a certain area

Examination-style

1 The graph shows the distribution of root lengths in a population of a species of grass.

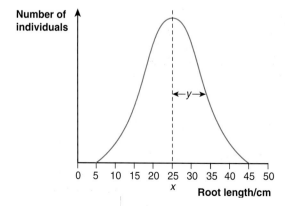

(a) (i) What does the term population mean? *(1 mark)*
(ii) What is the range of root lengths of this population? *(1 mark)*
(iii) Name the features of the distribution labelled *x* and *y*. *(2 marks)*
(b) Copy the graph and add to it the distribution of root lengths from another population which has the same mean root length, but shows much less variation. *(2 marks)*

Total: 6 marks

2 (a) The seeds from several pea plants were collected and weighed. Their masses are shown in the table.

Mass/g	Number of seeds
< 1.0	0
1.1–1.5	1
1.6–2.0	3
2.1–2.5	7
2.6–3.0	11
3.1–3.5	5
3.6–4.0	2
> 4.0	0

(i) Plot a graph of these results. *(3 marks)*
(ii) What sort of variation is shown by these data? *(1 mark)*
(b) The seeds were germinated in a greenhouse and the heights of the plants at 42 days were measured. The results are shown in the graph.

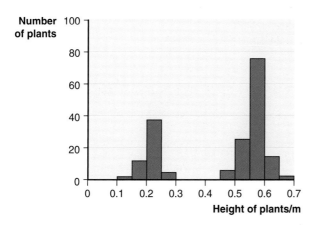

(i) Suggest why the seeds were germinated in a greenhouse rather than in the open. *(2 marks)*
(ii) Describe and explain the variation shown between the pea plants. *(4 marks)*

Total: 10 marks

3 Two populations of nettle plants were investigated. One population came from a shaded area, the other from a well-lit area. The surface area of the fifth leaf from the top of 20 plants chosen at random from each population was estimated. The results are shown in the table.

Surface area (cm^2) of leaves from the shaded site	Surface area (cm^2) of leaves from the well-lit site
20	18
15	19
13	12
23	14
26	13
19	16
18	14
21	12
17	15
19	17
21	14
22	12
25	12
19	16
16	17
21	14
24	15
27	11
22	16
18	13

Mean	20.3	
S.D.	3.66	2.24

(a) (i) Why were the plants chosen at random? (*1 mark*)

(ii) Describe how you could obtain a random sample of nettle plants from each site. (*2 marks*)

(b) (i) Calculate the mean for the well-lit site. Show your working. (*2 marks*)

(ii) Which sample shows the greater variability? Explain your answer. (*2 marks*)

Total: 7 marks

4 Scientists had observed that levels of plasma cholesterol and other lipids in women increase after the menopause (the period during which female fertility declines to zero). However, they were uncertain as to whether this was due to genetic or environmental factors. They decided to investigate by means of twin studies.

(a) In an investigation, they measured cholesterol levels in sets of identical and non-identical twins in each of three conditions:
- both twins pre-menopausal
- one twin pre-menopausal and one twin post-menopausal
- both twins post-menopausal

They then calculated concordance between cholesterol levels of the twins in each of the conditions. The concordance values are shown in the table:

	Both twins pre-menopausal	One twin pre-menopausal; one twin post-menopausal	Both twins post-menopausal	Overall
Identical	0.51	0.18	0.63	0.61
Non-identical	0.42	0.37	0.33	0.38

(i) Why were both identical and non-identical twins studied? *(2 marks)*

(ii) Why were the three menopausal condition pairs used? *(3 marks)*

(iii) Describe the differences in the patterns of the concordance of identical and non-identical twins. *(3 marks)*

(iv) What do the overall concordance values for identical and non-identical twins suggest about the cause of the change in cholesterol levels at the menopause. Explain your answer. *(3 marks)*

(v) Explain why these findings should be treated with caution. *(2 marks)*

(b) In another investigation, they decided to measure the heritability of plasma levels of another lipid. Heritability is given by the formula:

$$\text{heritability} = \frac{\text{variation due to genetic factors}}{\text{variation due to genetic factors} + \text{variation due to environment}}$$

The heritability for this lipid was 0.82. Describe the extent of the influence of the environment on plasma levels of this lipid. *(2 marks)*

Total: 15 marks

Chapter 10

DNA is the stuff of genes

This chapter covers:
- the work leading to the discovery that DNA is the hereditary material
- the structure of DNA
- DNA and chromosomes
- how genes determine characteristics
- how genes code for proteins
- the fact that not all DNA is coding DNA
- the consequence of changes in the genetic code

We are all familiar with the term 'DNA'. We have heard of DNA fingerprinting and the national DNA database. Some of us even know that DNA stands for deoxyribonucleic acid. Because it is so much in the news, we are aware that it is the molecule of which our genes are made. In this chapter we look at the structure of DNA and how that structure enables it to control activities in a cell, as well as ensuring that the DNA is passed, unaltered, to the next generation. We shall also see that most of our DNA is 'junk' and that very little of it is actually 'the stuff of genes'.

Box 10.1   How do we know that DNA is the hereditary material?

The evidence that DNA is the material of inheritance accumulated slowly. As often happens in scientific research, results from one experiment suggested new ideas to other researchers. Some of the key events were:

- In 1869, Johann Miescher, through chemical analysis, identified nucleic acids in cells, but not their location.
- In 1875, Eduard Strasburger described chromosomes and their location in the nucleus of a cell.
- In 1885, Albrecht Kossel followed up the research of Miescher and identified the sugar and organic bases in nucleic acids.
- In 1928, Frederick Griffiths showed that cell contents from a heat-killed pathogenic bacterium could transform a harmless strain of the same species of bacterium into the pathogenic strain.
- In 1944, Avery and MacLeod repeated Griffiths's experiment, but by treating the cell contents in a variety of ways were able to show that it was DNA that brought about the transformation.
- In 1952, Alfred Hershey and Martha Chase demonstrated that only the DNA of a virus is necessary to make a cell produce new viruses. DNA must be the material of inheritance.

The Hershey–Chase experiment

Hershey and Chase knew of the experiments by Griffiths and those of Avery and MacLeod. They devised a way of showing that DNA is passed from generation to generation and is responsible for 'directing' the synthesis of proteins. They worked with bacteriophages (literally 'bacteria eaters'). Bacteriophages are viruses that infect bacteria. They consist of a DNA 'core' surrounded by a protein coat. When a bacteriophage infects a bacterium, the DNA core is injected into the bacterial cell but the protein coat remains outside. Inside the cell, the viral DNA replicates itself and 'directs' the synthesis of more viral protein. The protein and DNA are assembled into new virus particles. This was the basis of their experiment. They needed to show that *only* the DNA of the virus directs the synthesis of protein.

Their investigation was as follows:

- Bacteriophages were labelled with radioactive sulphur and radioactive phosphorus. Sulphur is found in proteins but not in DNA; phosphorus is found in DNA but not in proteins. Therefore, the viruses contained DNA labelled with radioactive phosphorus and proteins labelled with radioactive sulphur.
- These bacteriophages were used to infect bacteria.
- The mixture was agitated in a blender in order to remove the viral coats.
- On filtering the mixture they found:
  - protein containing radioactive sulphur but no bacterial cells in the liquid
  - live bacterial cells containing radioactive phosphorus in the sediment
  This meant that only the DNA had entered the bacterial cells.
- On re-culturing the bacteria, the bacteriophages replicated themselves as normal.
- When the bacteriophages emerged from the bacterial cells, they contained DNA labelled with radioactive phosphorus.

This experiment showed that replication of the viruses could occur without the protein coat. The viral DNA could replicate itself, instruct the host cell to synthesise viral DNA and protein, *and* organise the DNA and viral protein into new virus particles. It provided strong evidence that DNA is the hereditary material.

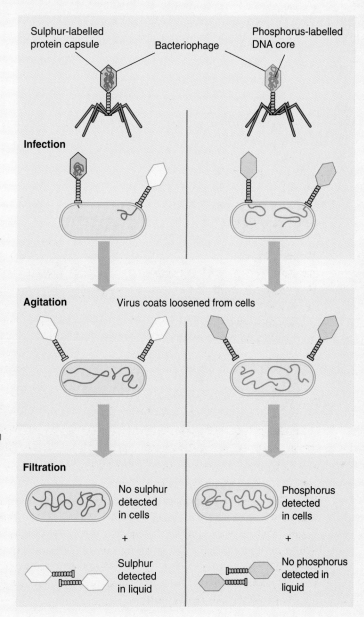

Sulphur-labelled protein capsule   Bacteriophage   Phosphorus-labelled DNA core

Infection

Agitation   Virus coats loosened from cells

Filtration

No sulphur detected in cells

+

Sulphur detected in liquid

Phosphorus detected in cells

+

No phosphorus detected in liquid

Box 10.2 How was the structure of DNA discovered?

The double helix of the DNA molecule is perhaps the most well-known of all molecular structures. The structure was unravelled in 1953 by two young research biochemists, James Watson and Francis Crick, and proved to be one of the greatest discoveries of all time. However, their work depended on that of others before them and, in particular, on that of Rosalind Franklin.

Rosalind Franklin provided X-ray diffraction photographs of DNA. Her colleague, Maurice Wilkins, suggested a spiral structure for the molecule.

Watson and Crick had the key elements from which to work out the structure of DNA. They knew:

- the nature of the sugars and the organic bases in the molecule (from Kossel's work in 1885)
- that the amounts of adenine and thymine are equal, as are the amounts of cytosine and guanine (discovered by Erwin Chargaff in the late 1940s)
- the molecule had a spiral structure (from Rosalind Franklin's work)

X-ray diffraction pattern of DNA, similar to that obtained by Rosalind Franklin

Using these ideas and a considerable amount of inspired guesswork Watson and Crick constructed the double-helix model of DNA.

Watson and Crick with their original model of DNA

What is DNA like?

DNA molecules are huge. Each molecule consists of two strands twisted into a double helix. Each strand is made from millions of subunits called **nucleotides** (Figure 10.1) and is, therefore, a polynucleotide. Each nucleotide has three components:

- an **organic base** — either adenine, thymine, cytosine or guanine
- the pentose sugar **deoxyribose**
- a **phosphate** group

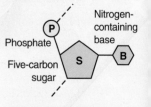

Figure 10.1 Structure of a nucleotide

There are four different bases, each containing nitrogen, so there are four different types of nucleotide in a DNA molecule.

Two nucleotides can react to form a dincleotide in a similar way to two monosaccharides reacting to form a disaccharide (Figure 10.2).

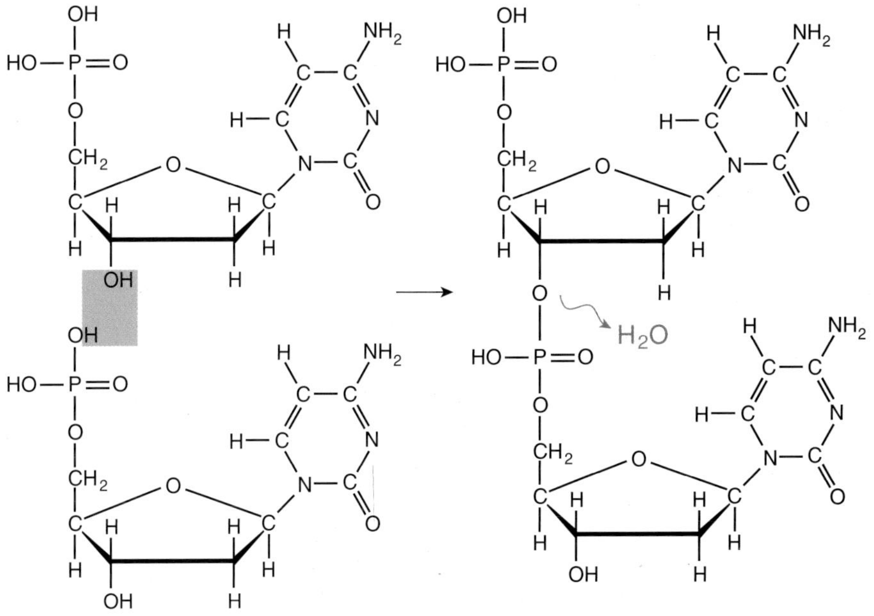

◀ An organic base is a base because each nitrogen atom has a lone pair of electrons that can attract a hydrogen ion. Overall, the molecule of DNA (deoxyribonucleic *acid*) is acidic because of the many phosphate groups it contains. Each –OH in a phosphate group can release a hydrogen ion, making the surrounding medium more acidic.

Figure 10.2 Two nucleotides reacting to form a dinucleotide; this is a condensation reaction

Be quite clear in your mind about the distinction between a *molecule* of DNA and a *strand* of DNA. There are *two* strands in *each* molecule.

The nucleotides in each strand of DNA are held together by bonds between the deoxyribose (pentose sugar) of one nucleotide and the phosphate group of the next. There is a 'sugar–phosphate backbone' to which the bases are attached (Figure 10.3).

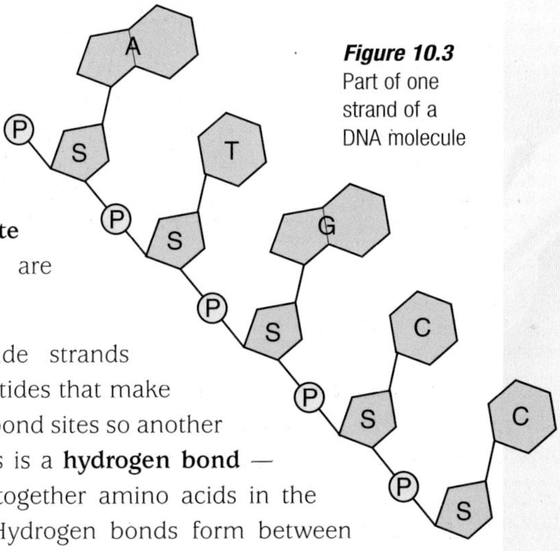

Figure 10.3 Part of one strand of a DNA molecule

What holds the two polynucleotide strands together? All the atoms in the nucleotides that make up each chain have no free covalent bond sites so another type of bond must be involved. This is a **hydrogen bond** — the same type of bond that holds together amino acids in the secondary structure of a protein. Hydrogen bonds form between the organic bases of each polynucleotide strand (Figure 10.4).

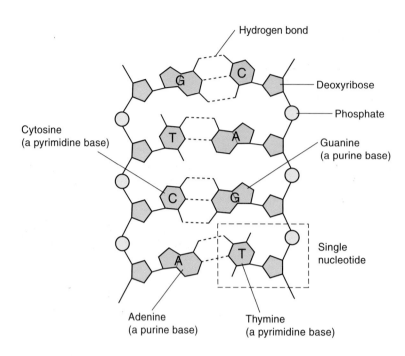

Hydrogen bond

Deoxyribose

Phosphate

Cytosine
(a pyrimidine base)

Guanine
(a purine base)

Single
nucleotide

Adenine
(a purine base)

Thymine
(a pyrimidine base)

Figure 10.4 The two
strands that make up a DNA
molecule are held together
by hydrogen bonds

◄ Adenine (A) always
bonds with thymine (T)
and guanine (G) always
bonds with cytosine (C).

There are two other important features concerning the way in which the two strands of DNA are held together:

- A nucleotide containing the base adenine on one of the strands is *always* paired with one containing thymine on the other strand. A nucleotide containing the base cytosine is *always* paired with one containing guanine. This is the **base-pairing rule**. It explains why, in 1950, Chargaff always found equal amounts of adenine and thymine in a molecule of DNA (whatever its origin), together with equal amounts of cytosine and guanine. Adenine and thymine are said to be **complementary bases**, as are cytosine and guanine.

- The strands are oriented opposite to each other. The 'start' or 'top' of one strand is paired with the 'end' or 'bottom' of the other. The two strands are said to be **anti-parallel**.

The percentage of adenine in a molecule of DNA is always equal to that of thymine and the percentage of cytosine is always equal to that of guanine. So, if just one of these figures is known, all the others can be calculated.

Suppose the percentage of cytosine in a molecule of DNA is 3 4%. By the base-pairing rule, the percentage of guanine must also be 34%. Cytosine and guanine together account for 68% of the bases in the molecule of DNA. Adenine and thymine must therefore account for 32% of the DNA. The amounts of adenine and thymine are equal, so each must account for 16%.

Box 10.3 Size and stability of DNA

DNA is a huge molecule, but its size varies from species to species. Describing this variation in size by using the difference in molecular mass gives some extremely large and rather unhelpful numbers. Instead, biologists use the idea that DNA is made from nucleotides containing bases — bigger molecules of DNA will contain more nucleotides and more bases. The size of a DNA molecule is expressed in terms of how many base pairs it contains. Because there are so many, these units are often kilobase pairs (thousands of base pairs) or megabase pairs (millions of base pairs).

DNA is a very stable molecule. If it were not, and were constantly changing, mutations would occur frequently. If it broke down easily, it would not be able to replicate effectively and the DNA passed on would not be identical to that in the original cell.

Where is DNA found?

We know from its name (**d**eoxyribo**n**ucleic **a**cid) that it is found in the nucleus. However, DNA does not float free in the nucleus; it is associated with proteins called **histones** and forms **chromosomes**.

Box 10.4 The structure of chromosomes

Chromosomes are composed of DNA and proteins called histones. The DNA–histone complex is called chromatin. When a cell is not dividing, the chromatin is loosely organised throughout the nucleus as loops of chromatin fibres. Individual chromosomes cannot be distinguished. The 'loose' organisation allows the genes to be active. As a cell prepares to divide, the chromatin loops (by now duplicated) become compacted to form a chromosome that is visible (when stained) under a light microscope. This stage, when the nucleus is preparing to divide by mitosis (see Chapter 11), is known as metaphase. The compact state of the chromatin in a chromosome means that the genes are too tightly packed to be active.

The different levels of organisation in a metaphase chromosome are shown in the diagram.

This organisation of DNA with histones into chromosomes is found in all eukaryotic cells.

The DNA of prokaryotic cells, however, is different from that of eukaryotic cells:

- it is much smaller
- it is circular, not linear
- it is not associated with histones to form chromosomes

What are genes and what do they do?

Genes are sections of DNA that determine characteristics. How do they do this? Each gene carries the code for the production of a particular protein. This could be a structural protein in the cell membrane or elsewhere in the cell, or an enzyme that controls reactions in the cell. Structural proteins and enzymes both contribute to the nature of a cell; each cell interacts with other cells to contribute to the properties of the whole organism, as shown in Figure 10.5.

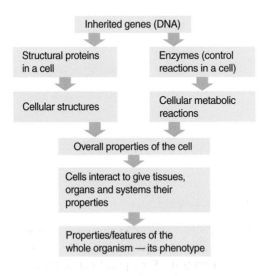

Figure 10.5 Genes affect the whole organism

Proteins are polymers of amino acids. Each protein has a unique tertiary structure because of the sequence of amino acids in its molecule. To control protein synthesis, DNA must be able to specify this sequence of amino acids. Each gene controls the synthesis of just one protein. This provides us with a useful working definition of a gene:

> A gene is a sequence of nucleotides at a fixed position on a strand of DNA that specifies (codes for) a sequence of amino acids that forms the primary structure of a protein.

◀ The position of a gene is called its **locus**.

Notice that we define a gene as a sequence of nucleotides on just *one* strand of DNA. This is called the **coding strand**. The other strand carries different sequences and is the **non-coding strand**.

As the sugar molecule and phosphate group are identical in all DNA nucleotides, it must be the sequence of bases in the gene that specifies the sequence of amino acids. The code for a protein that is specified by DNA has to be carried to the ribosomes so that they can assemble the amino acids in the correct sequence. However, DNA remains in the nucleus at all times. The following events occur in order to take the genetic code out of the nucleus, and convert it into an amino acid sequence:

◀ The coding strand is sometimes called the 'template' strand.

- The DNA code is 'rewritten' in a molecule of **messenger RNA** (mRNA) that travels from the nucleus through pores in the nuclear envelope to the ribosomes. This rewriting of the code is called **transcription**.

- Free amino acids are transferred from the cytoplasm to the ribosomes. This is carried out by molecules of **transfer RNA** (tRNA).
- The mRNA code is 'read' and the amino acids are assembled into a protein. This is called **translation** and is carried out by the ribosomes.

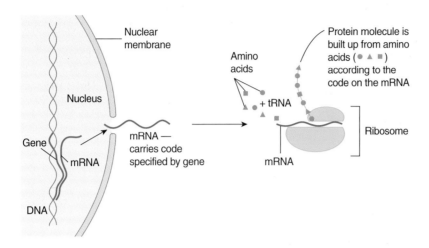

Figure 10.6 How DNA controls protein synthesis

How do genes code for proteins?

All the different proteins that are found in all the organisms on Earth are synthesised from just 20 amino acids. Different numbers and sequences of these amino acids produce an almost limitless range of proteins. Each protein is coded for by bases in a section of DNA, but there are only four different bases available. How can this be? If each base coded for a single amino acid, then it would only be possible to specify four different amino acids. If two bases were needed to code for one amino acid, then 16 amino acids could be specified. This is still not enough. A sequence of three bases per amino acid could code for 64 different amino acids. This is more than enough and is, in fact, the way in which the bases are organised to code for amino acids. The genetic code is a **triplet code**. It is also a **non-overlapping code**. This means that each triplet is distinct from all other triplets.

Hair contains the protein keratin. Bone contains the protein collagen. The protein in feathers is almost the same as that in the scales of dinosaurs.

A triplet of bases codes for each amino acid. There are 64 codes and only 20 amino acids, so what is the purpose of the other 44 codes? In fact, none of the codes is spare or redundant. Most amino acids have more than one code. Only methionine and tryptophan have just one triplet that codes for them; arginine has six. Three of the triplets (TAA, TAG and TGA) do not code for amino acids. They are 'stop' codes that signify the end of a coding sequence. Because there is this 'spare capacity' in the genetic code over and above what is essential, it is called a **degenerate code**.

The genetic code is also a **universal code**. For example, the triplet TAT is the DNA code for the amino acid tyrosine in a human, a giant redwood tree, an *E. coli* bacterium or in any other living organism.

The DNA codes for all 20 amino acids are given in Table 10.1.

Table 10.1

First position		Second position				Third position
		T	C	A	G	
T		Phenylalanine Phenylalanine Leucine Leucine	Serine Serine Serine Serine	Tyrosine Tyrosine stop stop	Cysteine Cysteine stop Tryptophan	T C A G
C		Leucine Leucine Leucine Leucine	Proline Proline Proline Proline	Histidine Histidine Glutamine Glutamine	Arginine Arginine Arginine Arginine	T C A G
A		Isoleucine Isoleucine Isoleucine Methionine	Threonine Threonine Threonine Threonine	Asparagine Asparagine Lysine Lysine	Serine Serine Arginine Arginine	T C A G
G		Valine Valine Valine Valine	Alanine Alanine Alanine Alanine	Aspartic acid Aspartic acid Glutamic acid Glutamic acid	Glycine Glycine Glycine Glycine	T C A G

In Table 10.1, the first letter of each triplet specifies a horizontal band. The second letter specifies a column and the third letter specifies a horizontal line. Take, for example, the triplet CAG:

- **C** (the first position) specifies the second horizontal band across
- **A** (the second position) specifies the third column
- **G** (the third position) specifies the fourth horizontal line

To find which amino acid is coded for by CAG find the fourth line in the third column of the second band. CAG is the code for the amino acid glutamine. ATT is the code for isoleucine and GAG is the code for glutamic acid. Check the table and see if you agree!

So the sequence of bases in a gene determines the sequence of amino acids?

Yes and no! We now know that not all DNA is 'coding DNA'. In humans only about 1% of DNA is actually coding sequence. There are sequences of bases *within* a gene called **introns** that do not code for any of the amino acids in the final protein. The coding regions are called **exons**. When the DNA is transcribed to mRNA, the transcribed mRNA introns are enzymically 'cut out'.

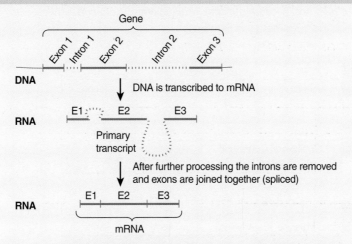
There are also base sequences *between* genes that do not code for amino acids. These sequences are often repeated These repeating sections are called **minisatellites** or **microsatellites** (smaller sections with fewer repeats). Figure 10.7 shows the base sequence of a minisatellite from a giant panda in which the base sequence CAA is repeated 12 times.

Figure 10.7 A minisatellite of a giant panda

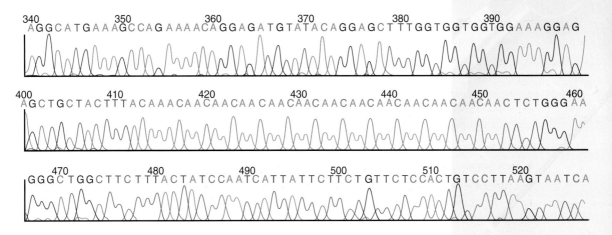

There are large numbers of microsatellites and minisatellites in the human genome; they vary between individuals. The variability of minisatellite DNA and microsatellite DNA is the basis of genetic fingerprinting. It is extremely unlikely that two individuals will have the same versions of several minisatellites or microsatellites.

What happens if the base sequence in a gene changes?

If the base sequence changes, the new sequence may code for a different sequence of amino acids — or it may not. Because the code is degenerate, it depends where the change occurs. Look at the examples in Figure 10.8.

Original code	TTA	CGG	ATC	TCC
Amino acids	Leucine	Arginine	Isoleucine	Serine

Changed code	TTA	CGG	ATC	TCG
Amino acids	Leucine	Arginine	Isoleucine	Serine
	NO CHANGE TO SEQUENCE			

Changed code	TTC	CGG	ATC	TCC
Amino acids	Phenylalanine	Arginine	Isoleucine	Serine
	CHANGED SEQUENCE			

Figure 10.8 Possible effects of a change in base sequence

◀ Changes in the DNA sequence are **mutations**. Changes in a single base, such as the ones shown are **point mutations**.

If the change produces a new amino acid sequence, then a new tertiary structure results. This may be very different from that of the original protein and, therefore, the protein may no longer be able to carry out its function. If the protein is an enzyme, it might no longer be able to catalyse its reaction, which may result in an entire metabolic pathway being blocked. This is illustrated in Figure 10.9.

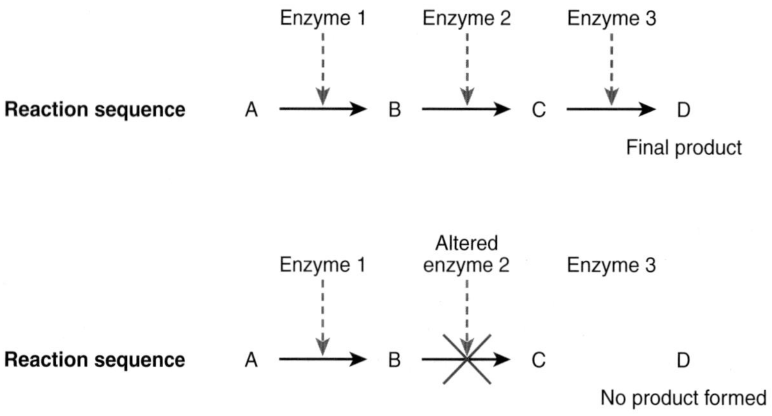

Enzyme 1   Enzyme 2   Enzyme 3

Reaction sequence   A ⟶ B ⟶ C ⟶ D
                                          Final product

                    Enzyme 1   Altered
                                enzyme 2   Enzyme 3

Reaction sequence   A ⟶ B ⤫ C   D
                                          No product formed

Figure 10.9 If a gene coding for an enzyme mutates, it may disrupt a whole metabolic pathway

Summary

Evidence that DNA is the hereditary material

- The Hershey–Chase experiment showed that DNA is the genetic material because it showed that DNA alone could 'instruct' cells to manufacture new viruses.
- In 1953 — using X-ray crystallography evidence supplied by Rosalind Franklin and ideas from Maurice Wilkins — James Watson and Francis Crick proposed the double-helix model for the structure of DNA.

The structure of DNA

The DNA molecule is a double helix in which:
- each strand is a polynucleotide consisting of four types of nucleotide
- each nucleotide contains the pentose sugar deoxyribose, a phosphate group and an organic base
- the four organic bases are adenine, thymine, cytosine and guanine
- specific base pairing occurs between the strands; adenine is always paired with thymine and cytosine with guanine
- each polynucleotide strand is held together by the 'sugar–phosphate backbone'; condensation links form between the phosphate group of one nucleotide and the deoxyribose sugar of the next
- the two strands are held together by hydrogen bonds
- only one strand of DNA holds coding information. This is the coding strand; the other strand is the non-coding strand

Genes

- A gene is a sequence of nucleotides at a fixed position on a strand of DNA. A gene specifies (codes for) a sequence of amino acids that forms the primary structure of a protein.
- The position of a gene is its locus.
- Genes code for the synthesis of proteins. They are transcribed to mRNA which travels to the ribosomes, where it is translated into a sequence of amino acids.
- The genetic code is:
  - triplet
  - non-overlapping
  - universal
  - degenerate
- Mutations in the genetic code can alter the protein produced; the new protein may be non-functional. If it is an enzyme it may lose its ability to catalyse its reaction; a whole metabolic pathway may be blocked as a result.
- Not all the bases in a molecule of DNA code for amino acids:
  - Introns are non-coding sequences within genes that are 'cut out' after transcription.

– Minisatellites and microsatellites are non-coding sequences found between genes, which contain repeats of a particular base sequence. They vary considerably between individuals and are used as the basis of genetic finger-printing.

Questions

Multiple choice

1 DNA consists of two polynucleotide strands in which:
   A the percentage of adenine is the same in each strand
   B the percentage of adenine is the same as that of thymine in each strand
   C the percentage of adenine is the same as that of thymine in the whole molecule
   D the percentage of adenine is 50 % of that of thymine in the whole molecule

2 The two strands of DNA are held together by:
   A hydrogen bonds
   B ionic bonds
   C covalent bonds
   D van der Waals forces

3 A nucleotide consists of:
   A an organic base, a pentose sugar and three phosphate groups
   B an inorganic base, a pentose sugar and three phosphate groups
   C an inorganic base, a pentose sugar and one phosphate group
   D an organic base, a pentose sugar and one phosphate group

4 Changes in the base sequence in a gene may not alter the protein produced because the genetic code is:
   A universal
   B degenerate
   C non-overlapping
   D a triplet code

5 The genetic code is:
   A a triplet code, degenerate and overlapping
   B a doublet code, degenerate and universal
   C a doublet code, degenerate and non-overlapping
   D a triplet code, degenerate and universal

6 Minisatellites contain:
   A repeating sequences of bases and are found between genes
   B repeating sequences of bases and are found within genes
   C non-repeating sequences of bases and are found between genes
   D non-repeating sequences of bases and are found within genes

7 A change in the sequence of bases in a gene may:
   A change the amino acid sequence but not affect the protein structure
   B not change the amino acid sequence but affect the protein structure
   C change the amino acid sequence and affect the protein structure
   D none of the above

8 The locus of a gene is:

 A the feature it codes for

 B its size

 C its base sequence

 D its position

9 Following transcription, mRNA must be modified to:

 A remove non-coding sections

 B alter its shape so that it can bind with ribosomes

 C alter its shape so that it can bind with an amino acid

 D remove unwanted amino acids

10 It is important that DNA is a stable molecule because:

 A it should not take part in any chemical reactions

 B alterations would result in mutations (changed genes) being passed on

 C transcription should always produce the same mRNA and, therefore, the same protein

 D all of the above

Examination-style

1 (a) The diagram represents the structure of the DNA molecule.

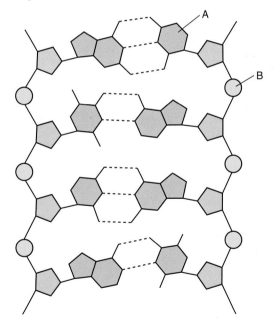

   (i) Name the structures labelled A and B.   (2 marks)

   (ii) The DNA molecule is sometimes described as consisting of two polynucleotide strands. Use the diagram to explain why. (1 mark)

 (b) Give three differences between DNA in prokaryotic cells and eukaryotic cells.   (3 marks)

 (c) In a molecule of DNA, adenine and thymine are present in equal amounts. Explain why.   (3 marks)

                                                        Total: 9 marks

2 The table below gives information concerning the percentages of the four different bases (adenine, thymine, cytosine and guanine) in the DNA of five different organisms.

Organism	Adenine/%	Guanine/%	Cytosine/%	Thymine/%
A	21	29	29	21
B	32	18	18	32
C			14	
D	13	37	37	13
E	26	24	24	26

(a) What percentages would you expect for adenine, guanine and thymine in the DNA of organism C? *(1 mark)*

(b) Explain why, in DNA, the ratio of the percentages of the bases adenine plus guanine to cytosine plus thymine (A + G:C + T) is always equal to 1. *(3 marks)*

(c) Explain why two organisms can have almost identical percentages of the four bases in their DNA and yet be very different organisms. *(2 marks)*

Total: 6 marks

3 The flow chart summarises the stages in the synthesis of proteins.

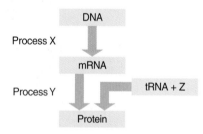

(a) (i) Name the processes taking place at X and at Y. *(2 marks)*

(ii) Name the substances represented by Z. *(1 mark)*

(b) Explain what introns are and why they must be removed during process X. *(3 marks)*

(c) Explain two potential consequences of a mutation of the base sequence AAA, which codes for the amino acid lysine. *(4 marks)*

Total: 10 marks

4 Minisatellites can be analysed to provide genetic fingerprints that can be used to determine parentage.

(a) What are minisatellites? *(3 marks)*

(b) Minisatellite DNA can be used as the basis for genetic fingerprints. Explain why. *(2 marks)*

(c) What are:

(i) introns? *(2 marks)*

(ii) exons? *(2 marks)*

Total: 9 marks

Chapter 11

Passing on DNA to the next generation

This chapter covers:
- DNA replication
- the cell cycle
- mitotic cell division
- mitosis and cancer
- meiotic cell division
- the role of mitosis and meiosis in life cycles

Multicellular organisms lose cells continually; these cells have to be replaced. These organisms also grow, which means that more cells are needed. Extra cells are produced only when existing cells divide.

When a cell divides to make two cells, the DNA in the original (parent) cell must be 'copied' into each of the two cells formed (the daughter cells). To produce this 'next generation of cells' the parent cell divides by **mitosis**.

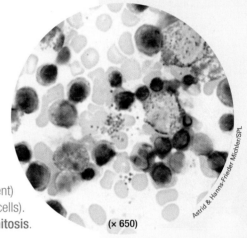

(× 650)

Red blood cells are produced by the mitotic division of cells in the bone marrow

However, to pass the DNA on to the next generation of *individuals* is not quite so straightforward. In sexual reproduction, two specialised sex cells fuse at fertilisation to form a single cell (the **zygote**) that begins the next generation. The sex cells must have only half the normal amount of DNA so that, when they fuse, the normal amount is restored. The cells that give rise to the sex cells divide by a different process, called **meiosis**.

◀ The cell theory put forward by Schleiden and Schwann (Chapter 2, page 25) says that cells can only arise from pre-existing cells.

How is DNA replicated?

When cells divide, it is important that the daughter cells formed (apart from the sex cells) contain the same genetic information as the parent cell that produced them. To achieve this, DNA must be able to replicate itself exactly.

DNA molecules exist within chromosomes in the nucleus and are surrounded by a 'soup' of free DNA nucleotides. On replication, it is these nucleotides that are used to build the new strands of DNA. Although they did not know the details,

Watson and Crick proposed that DNA replication would be **semi-conservative**. This means that the DNA molecule replicates in such a way that:

- each new DNA molecule formed contains one strand from the original DNA
- both new DNA molecules formed are identical to each other and to the original molecule

The process involves several enzymes and proteins, but the key stages are as follows:

- Molecules of the enzyme DNA helicase break hydrogen bonds and 'unwind' part of the helix of the DNA molecule, revealing two single-stranded regions.
- Molecules of DNA polymerase follow the helicase along each single-stranded region, which acts as a template for the synthesis of a new strand.
- The DNA polymerase assembles free DNA nucleotides into a new strand alongside each of the template strands. The base sequence in each of these new strands is complementary to its template strand because of the base-pairing rule: A–T, C–G (Chapter 10, page 170).
- The processes of unwinding followed by complementary strand synthesis progress along the whole length of the DNA molecule.
- The result is two DNA molecules that are identical to each other (and to the original molecule); each contains one strand from the original DNA molecule.

◀ The hydrogen bonds that hold the two strands of DNA together have approximately 5–10% of the strength of the bonds that hold the atoms together in the nucleotides. At room temperature, hydrogen bonds break more easily than other bonds. This means that the strands can be separated easily but the individual strands are stable.

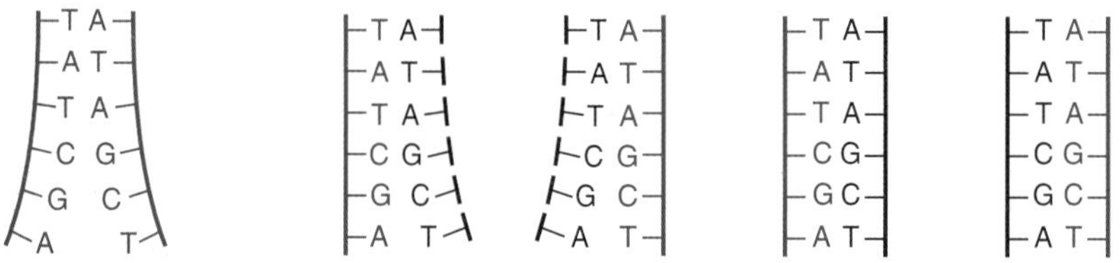

DNA helicase breaks the hydrogen bonds and the polynucleotide strands of DNA separate

Each strand acts as a template for the formation of a new molecule of DNA

Individual nucleotides line up with complementary bases on parent DNA strands

The nucleotides are joined together by DNA polymerase to form two molecules of DNA

Each molecule contains a strand from the parent DNA and a new strand

Figure 11.1 Semi-conservative replication of DNA

Although the idea of semi-conservative replication of DNA proposed by Watson and Crick was generally accepted as the likeliest method of replication, at the time there was no direct experimental evidence for it. In 1957, Matthew Meselson and Franklin Stahl carried out an investigation using the bacterium *Escherichia coli*, which provided strong evidence for the semi-conservative replication of DNA.

Box 11.1 The Meselson and Stahl experiment

Meselson and Stahl wanted to find a way of showing that when DNA replicates, one strand of the original molecule is 'conserved' in each of the two new 'daughter' molecules formed. After much thought, they proposed the following ideas:

- If the new DNA strands formed had slightly different masses from the original ones, this would affect the mass of the DNA molecules.
- Molecules of DNA with different masses could be identified using density-gradient centrifugation.
- If DNA replicates by semi-conservative replication, by creating conditions in which the DNA strands would have different masses, the positions in density-gradient centrifugation of the new strands could be predicted.

The basis of Meselson and Stahl's experiment is the technique of density-gradient centrifugation, which is very sensitive and can separate molecules of only slightly differing masses.

The element nitrogen has two isotopes. One isotope (the most common form) has an atomic mass of 14 (^{14}N). The other isotope has an atomic mass of 15 (^{15}N). *E. coli* bacteria grown in a culture medium in which all the nitrogenous compounds contained ^{15}N produced DNA that was slightly 'heavier' than 'normal' DNA containing ^{14}N.

Meselson and Stahl predicted that, if bacteria were grown in a medium containing only 'heavy' nitrogen, their DNA would all be 'heavy' DNA. If the bacteria were then transferred to a medium containing only 'light' nitrogen, as the DNA replicated, all the new strands would be 'light' DNA. According to semi-conservative replication, each molecule of DNA would then contain one heavy strand and one light strand; it would be intermediate in mass. When this intermediate DNA replicated (still in the 'light' nitrogen medium), all the new DNA

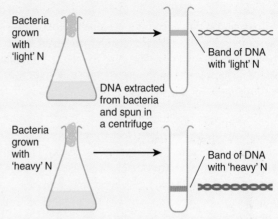

strands formed would be light DNA. However, half of these new strands would be paired with light strands from the intermediate molecules (forming light DNA) and half with the heavy strands from the intermediate molecules (forming more intermediate DNA). When the bacteria replicate again to form the third generation, those containing intermediate DNA will again produce 50% offspring containing intermediate DNA and 50% containing light DNA. Those containing light DNA will produce only offspring containing light DNA. Intermediate DNA now forms only 25% of the sample.

The experiment and results are summarised below.

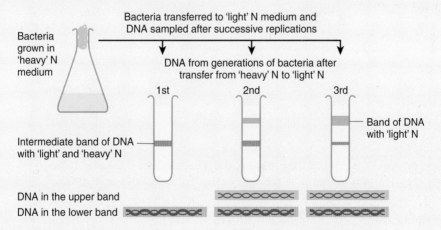

These results matched the predictions of Meselson and Stahl and provided strong evidence for the semi-conservative replication of DNA.

An understanding of the way in which DNA replicates paved the way for much of the research that was to underpin genetic engineering. Therefore, Meselson and Stahl's work was significant, not just for the scientific community, but for the community as a whole.

Cell factories and the cell cycle

Some cells in the human body retain their ability to divide by mitosis in order to replace those lost. Cells are continually being scraped off the lining of the gut as food passes through it. Tens of millions are lost each day from the stomach lining alone. About 1% of red blood cells die and are replaced each day — this amounts to a staggering 250 000 000 000 red blood cells. Cells are lost from the surface of the skin each time we touch something. These cells are also replaced by other cells dividing mitotically. However, there are some cells that have become so specialised that they have lost the ability to divide. For example, nerve cells do not normally divide, although some treatments can induce limited nerve re-growth.

Cells that divide repeatedly, such as stem cells in bone marrow, or cells in the skin that divide to replace those lost from the surface, go through a cycle of events called the **cell cycle**. These events eventually allow the cell to divide to form two cells by mitotic division. One of these cells eventually becomes a specialised cell; the other completes the cell cycle and replaces the original cell, so that the process can be repeated. The cell cycle is shown in the flowchart below, using a stem cell in bone marrow as the example.

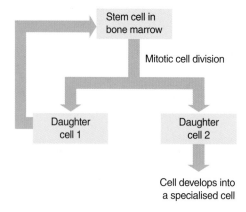

A cell that has just been produced by cell division must go through the following stages if it is to divide again:

- It must grow. Initially, the cell is half the size of the parent cell, with only half the organelles of a full-sized cell. During this phase of the cell cycle, more organelles are synthesised and the cell enlarges. Nucleotides and histone proteins are synthesised in preparation for DNA replication later in the cycle. This is the **G1 phase** of the cell cycle.
- DNA must replicate itself and combine with newly synthesised histone proteins to double the amount of chromatin in the nucleus. The cell continues to grow. This is the **S phase** of the cell cycle.
- The cell must prepare itself for mitosis. Specialised proteins called tubulins are synthesised. These are used to make the spindle apparatus, which will eventually separate the chromosomes. This is the **G2 phase** of the cell cycle.

◀ Adult fish can produce new brain cells — for example, the brown ghost fish can produce 50 000 per hour. It had been thought that cells in the human brain never divide, but recent research has shown that new brain cells are formed, albeit very slowly. This has opened up the possibility of future treatments for degenerative nerve conditions.

◀ Chromatin is the name given to the DNA–histone complex that makes up chromosomes.

The G1, S and G2 phases are collectively known as **interphase**. The nucleus of the cell now divides by mitosis. Once mitosis is complete, the cell divides into two cells by **cytokinesis**.

The DNA content of a cell changes during the cell cycle. Replication of DNA in the S phase means that the amount doubles. The amount of DNA remains at this 'double' level until cytokinesis because, until then, the DNA is still contained within one cell.

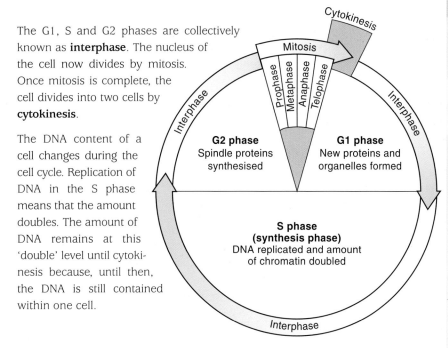

Figure 11.2 The cell cycle

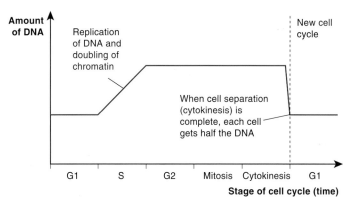

Figure 11.3 Changes in DNA content during the cell cycle

◄ In eukaryotic cells, DNA is combined with histones to form chromosomes. The behaviour of these chromosomes in cell division is responsible for DNA being passed to daughter cells.

Mitotic cell division

Mitotic cell division involves two main stages:
- division of the *nucleus* by mitosis
- division of the *cell* by cytokinesis

Mitosis

Mitosis results in each daughter nucleus containing exactly the same number and type of chromosomes as the parent nucleus from which they were formed. Each chromosome contains one molecule of DNA. Therefore, each daughter nucleus contains exactly the same amount and type of DNA as the other daughter nucleus and as the parent nucleus from which they were formed. This happens because, before mitosis, each molecule of DNA replicates and the chromosome that was a single structure becomes a double structure. Each of the two structures is a **chromatid**; the two chromatids that make up one chromosome are called 'sister chromatids'. They are held together by a **centromere**.

The difference in chromosome structure before and after DNA replication is shown in Figure 11.4.

Box 11.2 The centromere

The centromere is a constricted region of the chromosome that is made from two components:

● a specific sequence of DNA bases that is not transcribed, but which is required later for the segregation of the chromatids

● a protein-based structure called a kinetochore to which spindle fibres attach

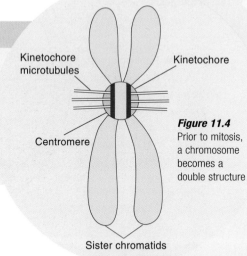

Kinetochore microtubules

Kinetochore

Centromere

Sister chromatids

Figure 11.4
Prior to mitosis, a chromosome becomes a double structure

Mitosis is a continuous process but there are four phases:
● prophase
● metaphase
● anaphase
● telophase

Box 11.3 How many cells are in each stage of the cell cycle?

Study the photograph of onion root tip cells. Most of the cells shown here are in interphase (no individual chromosomes visible). This is because growth of the cell, replication of DNA and production of the spindle proteins take longer than mitosis and cytokinesis. The approximate durations as a percentage of the total time for a cell cycle are:

● interphase — 80%, of which:
  – G1 phase = 25%
  – S phase = 25%
  – G2 phase = 30%
● mitosis — 17%, of which:
  – prophase = 6%
  – metaphase = 4%
  – anaphase = 3%
  – telophase = 4%
● cytokinesis — 3%

On a microscope slide showing 300 cells, one would expect to find 4% in metaphase, i.e. 12 cells.

The figures are approximations and there is considerable variation in the duration of cycles. However, the principle of calculating the number of cells in any one stage holds true.

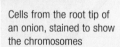

(× 1200)

Andrew Syred/SPL

Cells from the root tip of an onion, stained to show the chromosomes

Prophase

By the start of **prophase**, the chromosomes are double structures because, during the S phase of interphase, DNA has replicated. The two chromatids making up the chromosome are held together by a centromere. The chromatin becomes condensed during prophase, with the result that the chromosomes become shorter and thicker. The nucleolus disappears and the nuclear envelope begins to break down. Towards the end of prophase, **spindle fibres** appear.

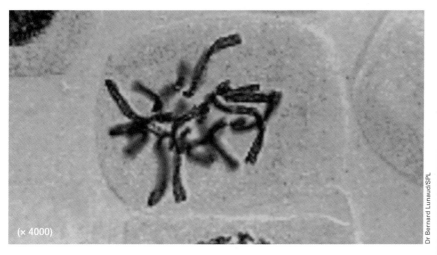

(× 4000)

Dr Bernard Lunaud/SPL

◀ The DNA in the seven pairs of chromosomes in a cell of the garden pea has a total length of about 7 m. By the end of prophase, the total chromosome length is just 350 μm. The length has been reduced by a factor of 20 000!

A bluebell cell during prophase of mitosis

Metaphase

During **metaphase**, the spindle apparatus develops fully. It contains two types of fibres:
- polar spindle fibres that extend from one pole of the cell to the other
- spindle fibres that extend from the pole of the cell and attach to the centromere of a chromatid

During metaphase, the chromosomes align along the centre of the spindle. The pattern of alignment determines how the chromatids will be segregated later.

Figure 11.5 The spindle apparatus

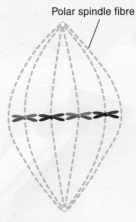

Polar spindle fibre

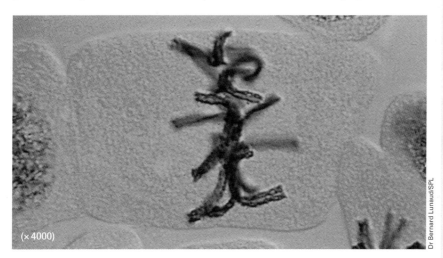

(× 4000)

Dr Bernard Lunaud/SPL

A bluebell cell in metaphase of mitosis

Anaphase

Anaphase is the shortest phase of mitosis. For each chromosome, it involves:

- the splitting of the centromere
- the separation of the two sister chromatids
- the movement of the sister chromatids to opposite poles of the cell

The spindle fibres pull the two chromatids from each chromosome to opposite poles of the cell. By the end of anaphase, there are two groups of chromosomes, one at each pole. Each group contains one 'chromatid' from each pair of sister chromatids.

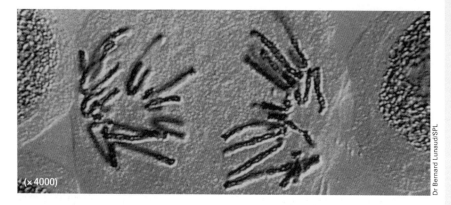

(× 4000)

Dr Bernard Lunaud/SPL

Telophase

During **telophase**, many of the events of prophase seem to be thrown into reverse:

- The spindle apparatus is dismantled.
- A nuclear envelope forms around each daughter nucleus.
- Nucleoli appear in each daughter nucleus.
- The chromosomes steadily elongate until there is, once again, just a diffuse mass of chromatin fibres.

By the end of telophase, the cell contains two daughter nuclei. These daughter nuclei are genetically identical because, before division, the sister chromatids in each chromosome contained identical DNA.

Telophase marks the end of mitosis. The original nucleus has divided into two genetically identical daughter nuclei. Next, the cell must divide into two.

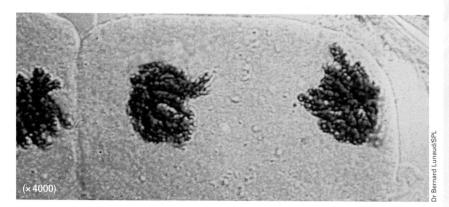

(× 4000)

Dr Bernard Lunaud/SPL

The mechanism by which the spindles pull the chromosomes apart remains uncertain. The most likely explanation is that the kinetochore shortens the spindle fibre to which it is attached by breaking off molecules of tubulin.

A bluebell cell in late anaphase of mitosis

Figure 11.6 Anaphase

Shortening spindle fibres

Polar spindle fibre

Chromatids separating

A bluebell cell in early telophase of mitosis

Cytokinesis

In animal cells, cytokinesis involves the plasma membrane forming a constriction across the centre of the cell. This becomes narrower and narrower, finally pinching off the cytoplasm into two cells. Plant cells must, in addition, synthesise two new cell walls across the centre of the original cell. Initially, a 'cell plate' is formed in the centre of the cell. This grows outwards and fuses with the cell wall, forming the two new cell walls and separating the two daughter plant cells.

Although the chromosomes are segregated in a precise manner during mitosis, the organelles of the original cell are distributed at random between the two daughter cells. Their position at the time of cytokinesis decides which daughter cell they enter.

Box 11.4 Mitotic cell division

The key features of a mitotic cell division are that:

- it involves only one nuclear division
- two daughter cells are formed
- the daughter cells are diploid (page 191)
- the daughter cells are genetically identical (contain the same number and type of chromosomes and therefore contain the same genes/DNA)

How are tumours formed?

Tumours form when a cell divides by mitosis in an uncontrolled fashion. The cells derived from this form a **clone** of genetically identical cells that also divide in an uncontrolled way. Very soon, a mass of cells called a tumour is formed. Most tumours are **benign**. These are usually slow growing, contained within a membrane and often harmless, although they may cause problems because of *where* they grow.

Malignant tumours divide more quickly and in a more uncontrolled way and are much more dangerous. It is these tumours that we call **cancers**.

A benign tumour in the brain may exert pressure on a region of the brain, causing damage to brain cells in that region. Or it may exert pressure on a blood vessel in the brain. This could either restrict the blood flow, or, possibly, cause the vessel to rupture.

Cancer is the Latin name for a crab. The extensions from the main tumour into the surrounding tissue sometimes look like the 'legs' of a crab — hence the name.

Box 11.5 Benign tumours and malignant tumours

Benign tumours and malignant tumours differ in a number of important ways:

- Benign tumours usually grow much more slowly than malignant tumours.
- Benign tumours usually remain encased within a fibrous capsule and do not invade the tissue in which they originated. The boundaries of malignant tumours are much less defined and the cells frequently invade the tissue in which they originate.
- Benign tumours rarely show **metastasis**, i.e. spread to other parts of the body. Many malignant tumours do metastasise and cause **secondary cancers**.

Cell division is usually regulated so that it takes place at the rate required. There are a number of control mechanisms that normally operate to prevent cell division from going out of control and forming tumours. **Proto-oncogenes** are

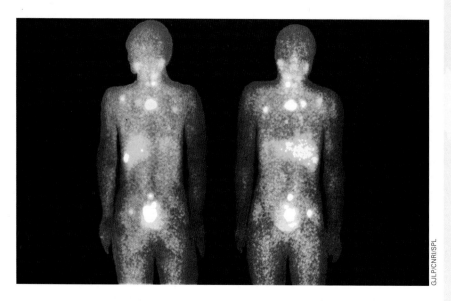

Coloured gamma scan of
a patient with metastatic
prostate cancer. The cancer
(white areas) has spread from
the prostate to several bones

inactive genes that are present in all our cells. They can be transformed into
active **oncogenes** in a number of ways. The gene may mutate, or it may be
influenced by a viral infection.

Active oncogenes produce proteins that interfere with the normal regulation
of cellular metabolism. The result is loss of control over cell division. In effect,
a switch is set that says 'keep dividing'. However, DNA that is damaged by
mutation is often repaired by the action of proteins produced by other genes.
Therefore, for a tumour to develop, this mechanism must fail too. A third set of
genes (**tumour-suppressor** genes) is also involved. These genes become active
when a group of cells is dividing in an uncontrolled manner. They 'switch off' the
division process. If they fail, a tumour forms. Some tumours are detected and
destroyed by our immune system. Those that are not may develop into cancers.

◀ Various environmental
factors increase the rate
of mutation of oncogenes.
These include **ionising
radiation** and specific
chemicals (cancer-makers
or **carcinogens**). Many
of the chemicals found
in tobacco smoke are
carcinogenic.

Figure 11.7 Development
of cancer

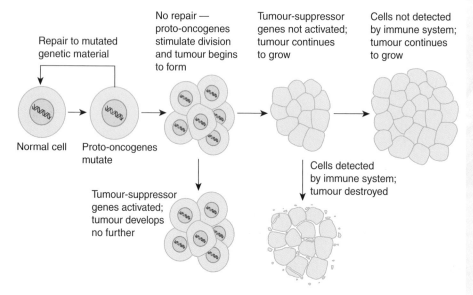

Repair to mutated
genetic material

No repair —
proto-oncogenes
stimulate division
and tumour begins
to form

Tumour-suppressor
genes not activated;
tumour continues
to grow

Cells not detected
by immune system;
tumour continues
to grow

Normal cell

Proto-oncogenes
mutate

Tumour-suppressor
genes activated;
tumour develops
no further

Cells detected
by immune system;
tumour destroyed

Meiotic cell division

Unlike mitosis, division of a nucleus by meiosis does not produce genetically identical daughter nuclei, but nuclei that show genetic variation. In mammals and higher plants, meiosis is involved only in the formation of sex cells.

Chromosome number

It is common knowledge that most human cells contain 46 chromosomes. However, there are some exceptions:

- Red blood cells contain none because they have no nucleus.
- Some liver cells contain 92.
- Sperm cells and oocytes contain 23.

The actual number of chromosomes does not tell the whole story. The 46 chromosomes in, for example, a skin cell, are in fact made up of 23 pairs — there are two sets of 23 different chromosomes. Each pair of chromosomes is called a **homologous pair**. Each chromosome within a homologous pair carries genes that control the same features in the same sequence. However, many genes can exist in two (or more) different forms called **alleles**. While the chromosomes of a homologous pair carry genes controlling the same feature, they may carry different alleles of those genes. However, the sister chromatids that make up one chromosome from the S phase of the cell cycle onwards are genetically identical and, therefore, carry the same alleles of the genes in the same sequence.

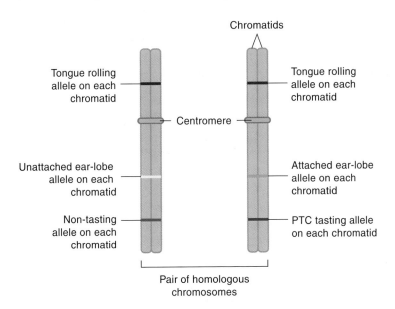

Pair of homologous chromosomes

Figure 11.8 Homologous chromosomes carry genes controlling the same feature in the same sequence; they might not carry the same alleles of those genes

Cells that contain two sets of chromosomes are called **diploid** cells. Human sperm cells and oocytes contain 23 chromosomes and therefore have only one set of chromosomes, i.e. they have one chromosome from each homologous pair. These cells are **haploid** cells.

Box 11.6 **Chromosome pairs**

Of the 23 pairs of chromosomes in most human cells, 22 pairs are autosomes (chromosomes not related to sex determination) and one pair are sex chromosomes. Cells from males have two different sex chromosomes, the X and the Y chromosomes. Cells from females have two X chromosomes.

The chromosomes are numbered by their size and shape. Of the autosomes, chromosome 1 is the largest and chromosome 22 is the smallest.

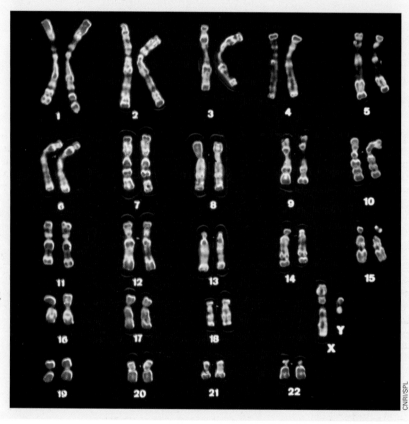

The 23 pairs of chromosomes from the cell of a human male

Different species have different numbers of chromosomes. Humans have 46 (two sets of 23), fruit flies have 8 (two sets of 4), garden peas have 14 (two sets of 7). The number of chromosomes in a 'set' is the **haploid number** and is represented by the letter n. The **diploid number** is therefore $2n$. So in humans, $2n = 46$.

Meiosis

As with mitotic cell division, meiotic cell division involves division of the nucleus (**meiosis**) and division of the cell (cytokinesis). However, in each full meiotic division, this happens twice to produce *four* daughter cells.

- **Meiosis I** separates the chromosomes from each homologous pair into different cells, reducing the chromosome number
- **Meiosis II** separates the chromatids in each chromosome, rather like mitosis

Figure 11.9 The main stages of meiosis I

Early prophase I
The DNA has already replicated and each chromosome consists of two chromatids. The cell contains two sets of chromosomes.

Late prophase I
The spindle starts to form. Homologous chromosomes pair up forming bivalents and exchange DNA between non-sister chromatids.

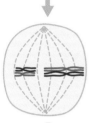

Metaphase I
The spindle is complete and the nuclear envelope has disintegrated. The bivalents are arranged around the middle of the spindle.

Anaphase I
Homologous chromosomes start to separate from each other.

Telophase I
There is one complete set of chromosomes at each end of the cell. The two cells resulting from meiosis I are therefore haploid.

Cytokinesis
The cytoplasm divides and two identical daughter cells are formed.

Box 11.7 Crossing over

Crossing over is a 'cut-and-paste' event that occurs when the two homologous chromosomes form a bivalent.

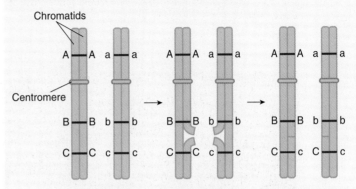

A pair of homologous chromosomes before crossing over. This shows the loci of 3 different genes and the different alleles of these genes on the chromosomes

During prophase 1 non-sister chromatids may break…

…and rejoin to the other chromosome. Now all four chromatids have different combinations of alleles — more variation exists

Box 11.8 Random segregation

The way in which the bivalents are aligned at metaphase, determines how they will be segregated into the two new cells.

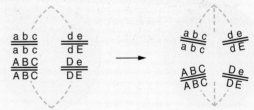

Sometimes the chromosome line up like this…

…giving this outcome

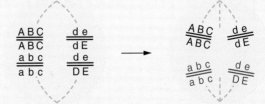

…and sometimes like this

…giving a different outcome

The main stages of meiosis II are shown in Figure 11.10.

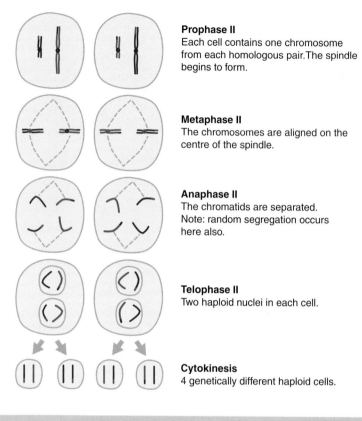

Prophase II
Each cell contains one chromosome from each homologous pair. The spindle begins to form.

Metaphase II
The chromosomes are aligned on the centre of the spindle.

Anaphase II
The chromatids are separated. Note: random segregation occurs here also.

Telophase II
Two haploid nuclei in each cell.

Cytokinesis
4 genetically different haploid cells.

Figure 11.10 The main stages of meiosis II

Box 11.9 Meiotic cell division

The key features of a meiotic cell division are that:

- it involves two nuclear divisions
- four daughter cells are formed
- the cells formed are haploid (page 191)
- the daughter cells show genetic variation (contain the same number and type of chromosomes, but different combinations of alleles)

Mitosis, meiosis, life cycles and chromosome numbers

Members of the same species have the same number and types of chromosome. The individuals vary because of the different combination of alleles that each possesses. In a species that reproduces sexually, mitosis and meiosis are important in keeping this chromosome number constant from one generation to the next. The roles of mitosis and meiosis in the life cycles of a mammal and of a flowering plant are shown in Figure 11.11.

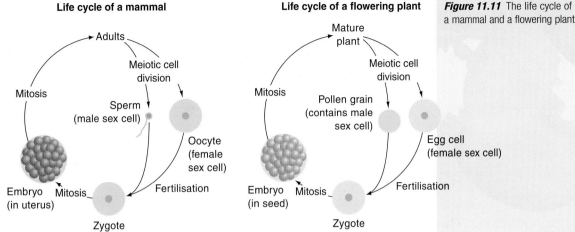

Life cycle of a mammal

Life cycle of a flowering plant

Figure 11.11 The life cycle of a mammal and a flowering plant

In both these life cycles:

- cells in the sex organs divide meiotically to form haploid sex cells
- fertilisation produces a zygote in which the normal diploid number of chromosomes has been restored
- from the zygote, mitotic cell division produces all the diploid cells of the adult organism

Summary

Why cells divide

- In mammals and higher plants, cells divide mitotically to allow growth to take place and to replace cells that are lost or damaged.
- In mammals and higher plants, cells divide meiotically to produce sex cells.

DNA replication

- DNA replicates itself as a precursor to cell division.
- During the replication of DNA:
  - DNA helicase breaks the hydrogen bonds to separate the two strands and 'untwist' the helix
  - each separated strand forms a template for the synthesis of a new complementary strand
  - using free DNA nucleotides, the enzyme DNA polymerase assembles a complementary strand for each of the separated strands of the original molecule
  - the new strand binds with the original strand by forming hydrogen bonds between the complementary bases
  - two complete molecules of DNA are formed that are identical to each other and to the original molecule of DNA from which they were formed
- DNA controls protein synthesis and cell activity

The cell cycle

- Stem cells go through the cell cycle repeatedly; other cells lose the ability to divide when they become specialised.
- The main phases of the cell cycle are:
  - the G1 phase — the cell grows and produces more organelles
  - the S phase — growth continues, DNA replicates and each chromosome is duplicated to become a pair of sister chromatids, held together by a centromere
  - the G2 phase — the cell prepares for mitosis by producing tubulins (the proteins needed to assemble the spindle fibres)
  - mitosis — the nucleus divides into two genetically identical daughter nuclei
  - cytokinesis — the cell splits into two daughter cells
- The G1, S and G2 phases are known collectively as interphase.
- Chromosomes are made of chromatin, which is a complex of DNA and histone (a protein).

Mitotic cell division

- There are four stages in mitosis:
  - prophase — chromatin condenses, making the chromosomes shorter and thicker; the nucleolus and nuclear envelope disappear and assembly of the spindle apparatus commences
  - metaphase — assembly of the spindle apparatus is completed and chromosomes attach to the centre of the spindle by their centromeres
  - anaphase — spindle fibres shorten, split the centromere and pull sister chromatids (now called chromosomes again) to opposite poles of the cell
  - telophase — the two groups of chromosomes form daughter nuclei; the spindle is dismantled, nuclear envelopes and nucleoli appear and the chromosomes become longer and thinner as the chromatin becomes less condensed
- Telophase is followed by cytokinesis, which involves:
  - in animal cells, the plasma membrane forming a constriction that eventually 'pinches off' the cytoplasm, forming two new cells
  - in plant cells, a cell plate forming in the centre of the cell, which grows outwards and forms two new cell walls that separate the daughter cells
- A single mitotic division produces two daughter cells that are diploid and genetically identical.

Mitosis and cancer

- Tumours are formed when cells divide by mitosis in an uncontrolled manner.
- Benign tumours:
  - are contained within a fibrous membrane
  - are composed of cells that divide slowly
  - do not invade surrounding tissues
  - do not metastasise (spread to other organs via the blood/lymph)

- Malignant tumours (cancers):
  - are not contained in a membrane
  - are composed of quickly dividing cells
  - invade the surrounding tissues
  - may metastasise
- Tumours form when:
  - proto-oncogenes mutate into oncogenes, which stimulate mitosis
  - repair of the mutated oncogenes fails
  - tumour-suppressor genes fail to restrict mitosis

Meiotic cell division

- In meiosis I:
  - crossing over takes place in prophase I, introducing genetic variation
  - the chromosomes in a homologous pair are separated at anaphase I
  - random segregation, which also introduces genetic variation, takes place at anaphase I,
  - two haploid cells are produced
- In meiosis II:
  - the chromatids in each chromosome are separated at anaphase II
  - random segregation takes place at metaphase II
  - four haploid cells showing genetic variation are produced

Questions

Multiple-choice

1 A mitotic cell division produces:
  - **A** two haploid cells
  - **B** two diploid cells
  - **C** four haploid cells
  - **D** four diploid cells
2 The sequence of phases in the cell cycle is:
  - **A** G1 → G2 → S → mitosis → cytokinesis
  - **B** G1 → G2 → mitosis → cytokinesis → S
  - **C** G1 → S → G2 → cytokinesis → mitosis
  - **D** G1 → S → G2 → mitosis → cytokinesis
3 A homologous pair of chromosomes is:
  - **A** a pair of sister chromatids held together by a centromere
  - **B** a pair of non-sister chromatids
  - **C** a pair of chromosomes carrying genes for the same features in the same positions
  - **D** a pair of chromosomes carrying genes for different features
4 During prophase:
  - **A** the nucleolus and nuclear envelope disappear and chromatin becomes much less condensed
  - **B** the nucleolus appears, the nuclear envelope disappears and chromatin condenses

C the nucleolus appears, the nuclear envelope appears and chromatin condenses

D the nucleolus and nuclear envelope disappear and chromatin condenses

5 The role of meiosis in life cycles is to:

   A restore the diploid chromosome number in the zygote

   B produce the haploid chromosome number in the zygote

   C produce the haploid chromosome number in the sex cells

   D produce the diploid chromosome number in the sex cells

6 During the S phase of the cell cycle, the DNA content of a cell:

   A doubles

   B halves

   C remains the same

   D none of the above

7 When the spindle apparatus is fully assembled:

   A all of the fibres are attached to chromosomes

   B all of the fibres run from one pole to the other

   C some fibres are attached to chromosomes, others run from one pole to another

   D all of the fibres are fully contracted

8 During cytokinesis in animal cells:

   A a cell plate is formed

   B new organelles are synthesised

   C DNA replicates

   D a constriction forms in the middle of the cell

9 Haploid cells have:

   A only one chromosome from each homologous pair

   B half the normal number of chromosomes for that species

   C both A and B

   D neither A nor B

10 In mammals, meiotic cell division differs from mitotic cell division in that:

   A the original nucleus is haploid, not diploid

   B the nucleus of each daughter cell is haploid, not diploid

   C there is only one nuclear division, not two

   D there are two daughter cells formed, not four

Examination-style

1 Figure 1 shows a cell in a stage of mitosis. The cell contains just two pairs of homologous chromosomes.

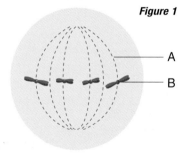

Figure 1

A

B

(a) (i) What are homologous chromosomes? *(1 mark)*
(ii) Identify the structures labelled A and B on Figure 1. *(2 marks)*
(iii) Name the stage of mitosis represented in this diagram.
Give a reason for your answer. *(1 mark)*

Figure 2 shows the life cycle of a mammal.

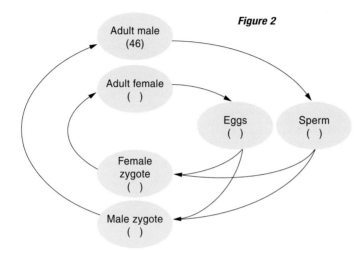

Figure 2

(b) (i) Copy Figure 2 and mark one stage where meiosis takes
place and one stage where mitosis takes place. *(2 marks)*
(ii) Complete the empty boxes to show the number of
chromosomes per cell. *(1 mark)*

Total: 7 marks

2 (a) The diagram shows the four stages of mitosis, but not necessarily in the
correct order.

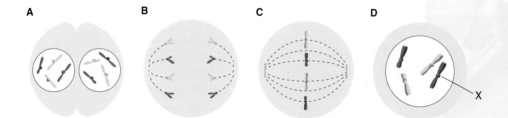

(i) Name the structure labelled X. *(1 mark)*
(ii) List the letters representing the stages in the order in which
the stages occur during mitosis. *(1 mark)*
(iii) Describe fully what happens during stage B. *(3 marks)*
(b) Describe three differences between mitosis and meiosis. *(3 marks)*

Total: 8 marks

3   The graph shows the changes in the amount of DNA in a cell during the stages
    of the cell cycle.

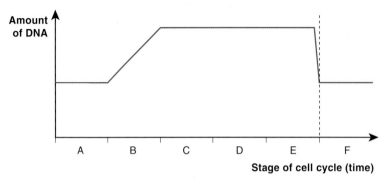

    Explain the change in the amount of DNA during:
    (a)(i) stage B   (2 marks)
       (ii) stage E   (2 marks)
    (b)(i) Name the stage of the cell cycle labelled C.   (1 mark)
       (ii) Describe the events that take place in the stage labelled A.   (2 marks)
                                                            Total: 7 marks

4   (a) Copy and complete the table below to compare mitosis and meiosis.

Feature of the process	Mitosis	Meiosis
Number of nuclear divisions		
Number of cells formed		
Genetic variability of daughter cells (Yes/No)		
Diploid or haploid daughter cells formed		

                                                                          (2 marks)

    (b) The graph shows the change in distance between the centromere of one
        chromosome and a pole of the cell.

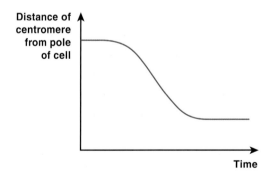

       (i) Describe the change in position of the chromosome.   (1 mark)
       (ii) Explain how this change in position is brought about.   (2 marks)
    (c) Explain why individual chromosomes are not visible when an
        interphase nucleus is viewed through a light microscope.   (2 marks)
                                                            Total: 7 marks

5 (a) The table below shows the relative amounts of DNA in some mammalian cells.

Cell	Relative amount of DNA/ arbitrary units
Skin cell	30
Muscle cell	30
Red blood cell	0
Sperm	
Oocyte	
Zygote	

(i) Give an explanation for the amount of DNA in a red
    blood cell.   (2 marks)

(ii) How much DNA would you expect there to be in:

   A a sperm

   B a zygote

   Explain your answers.   (4 marks)

(b) Some anticancer drugs inhibit spindle formation in mitosis.
    Suggest why patients given such drugs often develop anaemia.
    Hint: anaemia can be a result of too few red blood cells in
    the blood.   (3 marks)

Total: 9 marks

6 Scientists collected data on the most common cancers in men and women
   in the UK. The results are shown in the table.

Site (men)	% of all cancers (men)	Site (women)	% of all cancers (women)
Lung	21	Breast	19
Skin	14	Skin	11
Prostate	10	Lung	8
Bladder	5	Colon	6
Colon	6	Stomach	3
Stomach	5	Ovary	3
Rectum	5	Cervix	3
Lymph nodes	3	Rectum	3
Oesophagus	2	Uterus	2
Pancreas	2	Bladder	2
Other cancers	27	Other cancers	40

(a)(i) The data suggest that there are more cancers of the lung
       in men than in women. This is not necessarily the case.
       Explain why.   (2 marks)

(ii) Suggest and explain *two* reasons why the incidence of
     skin cancer appears to be higher in men than in women.   (4 marks)

(b) The tumour suppressor gene P53 is mutated in many human cancer cells. Scientists are trying to find ways of repairing the damaged gene. In one trial, they are using altered viruses to carry normal P53 genes into cancer cells. The treatment has shown promising results in laboratory animals.

(i) Why would introducing normal P53 genes into the cancer cells be an effective way of treating the cancer? *(3 marks)*

(ii) Give *scientific* reasons why the use of laboratory animals in this research may not be justified. *(2 marks)*

(iii) The use of viruses to deliver the P53 gene in humans may not be effective. Suggest *two* reasons why. *(4 marks)*

Total: 15 marks

Chapter 12

Different DNA results in different substances and different cells

This chapter covers
- how differences in DNA affect the biochemical composition of organisms
- the structure and function of haemoglobin and how this differs in different animals
- the differences in structural and storage carbohydrates in plant and animal cells
- some differences between plant and animal cells
- how different animal tissues arise
- how tissues are organised into organs and organs into organ systems

DNA is 'the stuff of our genes'. The genes code for a range of proteins, which have a variety of functions in cells. So different genes mean different proteins and cells with different functions. It is therefore fairly easy to appreciate why plant and animal cells are different. Different DNA results in different substances in the cells. For example, plant cells can make cellulose for their cell walls. But we also know that all our cells contain exactly the same genes (and so the same DNA) because they were produced by mitosis from a zygote — the single cell formed as a result of fertilisation. So why do we have such a range of different cells? And how are these different cells organised in our bodies?

Why does different DNA make different substances?

We need to think again about the role of DNA in a cell to answer this question:

Figure 12.1 The role of DNA in a cell

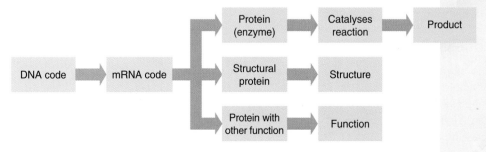

Different DNA results in different mRNA. This leads to the production of different proteins and:

- differences in the way a reaction is catalysed or different reactions being catalysed, *or*
- variations in a structure or different structures being produced, *or*
- variations in the function of a protein or proteins with different functions

One example of a protein with a specific function is haemoglobin.

What is haemoglobin and what does it do?

Haemoglobin is a protein with a quaternary structure. It is composed of four polypeptide chains, linked together to form a single molecule. Each polypeptide chain contains a 'haem' group', which contains iron. It is this haem group with which oxygen binds. There are four haem groups in each haemoglobin molecule, so each molecule can carry four molecules of oxygen.

It is the tertiary structure of each chain in the haemoglobin molecule that allows it to bind effectively with one oxygen molecule.

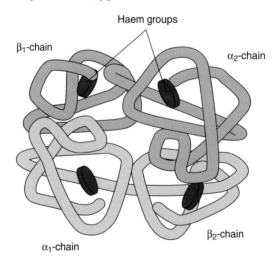

Haem groups

β_1-chain

α_2-chain

β_2-chain

α_1-chain

Figure 12.2 The quaternary structure of haemoglobin

Haemoglobin can bind (associate) with oxygen atoms where oxygen is plentiful in the surroundings (in the lungs) and release the oxygen (dissociate) when oxygen is scarce (in actively respiring tissues).

Loading and **unloading** oxygen are other terms sometimes used to describe the binding and releasing of oxygen ◄ by haemoglobin.

How does haemoglobin deliver oxygen to tissues?

Different concentrations of oxygen affect its binding to haemoglobin, but it is not an 'all-or-nothing' effect. Concentrations of respiratory gases are often expressed as **partial pressures** (Chapter 6) in units of kilopascals (kPa). The partial pressure of oxygen is often abbreviated to pO_2. Similarly, pCO_2 is an abbreviation for ◄ partial pressure of carbon dioxide.

The term **oxygen tension** means the same as partial ◄ pressure of oxygen.

The percentage of haemoglobin that is associated with oxygen at a given partial pressure of oxygen (pO_2) is referred to as the **percentage saturation** of haemoglobin. It varies with the oxygen tension of the surroundings. This is shown in a graph called the **dissociation curve** of haemoglobin.

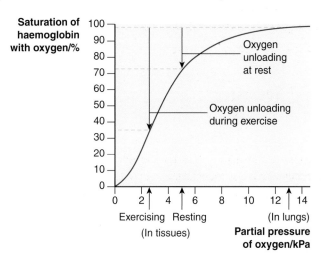

Figure 12.3 Oxygen dissociation curve of human haemoglobin

In the lungs, the partial pressure of oxygen is 13 kPa, and 98% of the haemoglobin associates (binds) with oxygen.

In respiring tissues at rest, the oxygen tension is lower, at 5.3 kPa. Under these conditions, only 73% of the haemoglobin is associated with oxygen. Therefore, 25% of the haemoglobin has unloaded its oxygen (released its oxygen to the tissues).

With moderate exercise, the oxygen tension in respiring muscle tissue falls even lower — to 2.5 kPa — and only 35% of the haemoglobin can remain bound to oxygen. Over 60% of the haemoglobin unloads its oxygen to meet the increased demand.

The association/dissociation of oxygen and haemoglobin is a reversible reaction:

$$HbO_2 \rightleftharpoons Hb + O_2$$

oxyhaemoglobin $\rightleftharpoons$ haemoglobin + oxygen

In respiring tissue, oxyhaemoglobin dissociates and unloads oxygen because:
- oxygen is rapidly used by respiring tissue and so diffuses rapidly from red blood cells, reducing the partial pressure of oxygen in the red cells
- the release of carbon dioxide by respiring cells leads to the production of hydrogen ions which bind with free haemoglobin, reducing the concentration of free haemoglobin in the red cells

An increased concentration of carbon dioxide shifts the dissociation curve to the right. The hydrogen ion concentration also increases (pH decreases) as a result. This results in increased unloading of oxygen at *any* partial pressure of oxygen. This effect is named after its discoverer and is called the **Bohr effect**.

In any reversible reaction, an equilibrium can be reached. Removal of the products (on the right-hand side of the equation) stimulates the reaction in that direction (to make more products). If the products accumulate, this inhibits the reaction ◀ in this direction.

Carbon dioxide dissolves in water to form carbonic acid. Like any acid, this releases hydrogen ions into the solution and causes a decrease ◀ in the pH of the solution.

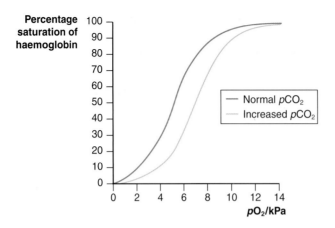

Figure 12.4 The Bohr effect

Haemoglobin is just haemoglobin, isn't it?

The haemoglobin molecule is similar in all animals that possess it, but there are differences. For example, the haemoglobin of the lamprey (a primitive fish-like animal) has only one polypeptide chain, not four. Most animals have haemoglobin with four chains, but the chains do vary. Figure 12.5 shows the differences in the sequence of amino acids in the α-chains of human and horse haemoglobin.

Figure 12.5 Human and horse haemoglobin have some different amino acids in the α-chains

Human	Val	Leu	Ser	Pro	Ala	Asp	Lys	Thr	Asp	Val	Lys	Ala	Ala	Try	Gly	Lys	Val	Gly	Ala	His	Ala	Gly	Glu	Tyr	Gly	Ala	Glu	Ala	Leu	Glu
Horse	Val	Leu	Ser	Ala	Ala	Asp	Lys	Thr	Asp	Val	Lys	Ala	Ala	Try	Ser	Lys	Val	Gly	Gly	His	Ala	Gly	Glu	Tyr	Gly	Ala	Glu	Ala	Leu	Glu
				5					10					15					20					25					30

Human	Arg	Met	Phe	Leu	Ser	Phe	Pro	Thr	Thr	Lys	Thr	Tyr	Phe	Pro	His	Phe	Asp	Leu	Ser	His	Gly	Ser	Ala	Glu	Val	Lys	Gly	His	Gly	Lys
Horse	Arg	Met	Phe	Leu	Gly	Phe	Pro	Thr	Thr	Lys	Thr	Tyr	Phe	Pro	His	Phe	Asp	Leu	Ser	His	Gly	Ser	Ala	Glu	Val	Lys	Ala	His	Gly	Lys
				35					40					45					50					55					60

Human	Lys	Val	Ala	Asp	Ala	Leu	Thr	Asp	Ala	Val	Ala	His	Val	Asp	Asp	Met	Pro	Asp	Ala	Leu	Ser	Ala	Leu	Ser	Asp	Leu	His	Ala	His	Lys
Horse	Lys	Val	Ala	Asp	Gly	Leu	Thr	Asp	Ala	Val	Gly	His	Leu	Asp	Asp	Leu	Pro	Gly	Ala	Leu	Ser	Ala	Leu	Ser	Asp	Leu	His	Ala	His	Lys
				65					70					75					80					85					90

Human	Leu	Arg	Val	Asp	Pro	Val	Asp	Phe	Lys	Leu	Leu	Ser	His	Cys	Leu	Leu	Val	Thr	Leu	Ala	Ala	His	Leu	Pro	Ala	Glu	Phe	Thr	Pro	Ala
Horse	Leu	Arg	Val	Asp	Pro	Val	Asp	Phe	Lys	Leu	Leu	Ser	His	Cys	Leu	Leu	Ser	Thr	Leu	Ala	Ala	His	Leu	Pro	Asp	Asp	Phe	Thr	Pro	Ala
				95					100					105					110					115					120

Human	Val	His	Ala	Ser	Leu	Asp	Lys	Phe	Leu	Ala	Ser	Val	Ser	Thr	Val	Leu	Thr	Ser	Lys	Tyr	Arg
Horse	Val	His	Ala	Ser	Leu	Asp	Lys	Phe	Leu	Ser	Ser	Val	Ser	Thr	Val	Leu	Thr	Ser	Lys	Tyr	Arg
				125					130					135					140	

Variations in amino acid sequences produce haemoglobin molecules with slightly different properties. This can be an evolutionary adaptation to environmental conditions. Some animals live in conditions in which the partial pressure of oxygen is much less than the 21 kPa of atmospheric air. These include:

- the llama, which lives at high altitudes (overall air pressure decreases with altitude and so the partial pressure of oxygen also decreases)

- *Arenicola*, a worm that lives in burrows in sand in the intertidal regions of the sea shore where partial pressure of oxygen is low
- the human fetus — the partial pressure of oxygen is much lower in the womb than in the lungs

For these animals to survive, their haemoglobin must show an increased affinity for oxygen. It must be able associate with oxygen at partial pressures where other haemoglobin molecules would not. The haemoglobin of the human fetus must associate with oxygen at partial pressures of oxygen at which the maternal oxyhaemoglobin *dissoc*iates.

Figure 12.6 The oxygen dissociation curves of a human fetus and adult

Why can plant cells make cell walls?

It is because the DNA in plant cells has some fundamental differences from that in animal cells. Plant DNA codes for enzymes that allow plants to polymerise β-glucose (a monosaccharide) into cellulose (a polysaccharide), from which the cell wall is made.

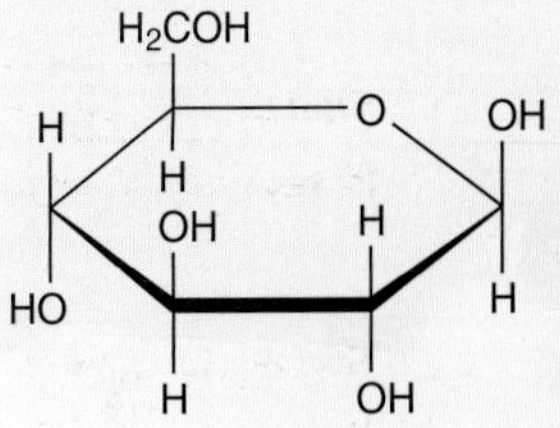

Figure 12.7 Structure of β-glucose

Two β-glucose molecules can be combined in a condensation reaction to form cellobiose:

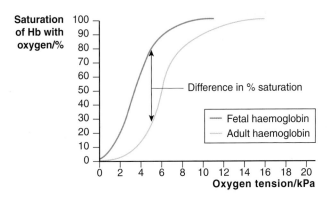

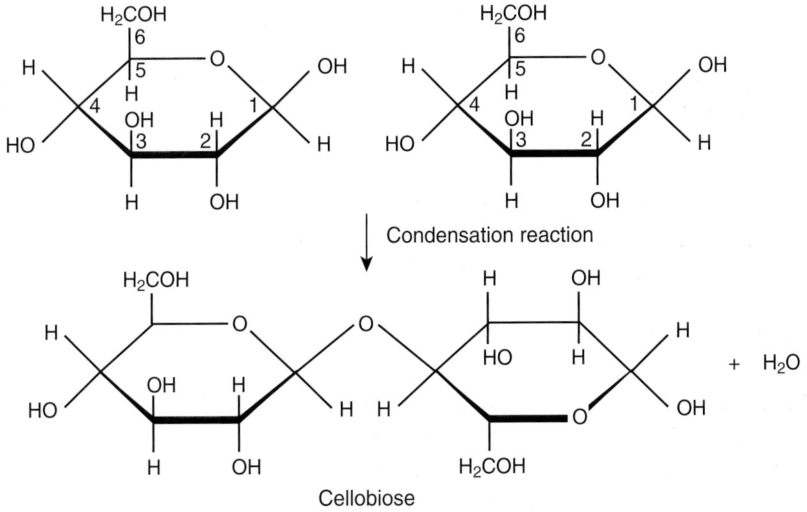

Figure 12.8 Formation of cellobiose

The bond formed is a β-1,4-glycosidic bond, as it forms between carbon atoms 1 and 4 of two β-glucose molecules.

Many β-glucose molecules link together in this way to form a molecule of cellulose. Because it is a linear molecule, with no branches, several cellulose molecules can hydrogen bond together to form a cellulose micelle. Several micelles are organised into a microfibril and several microfibrils make a cellulose fibre. This organisation is shown in Figure 12.9.

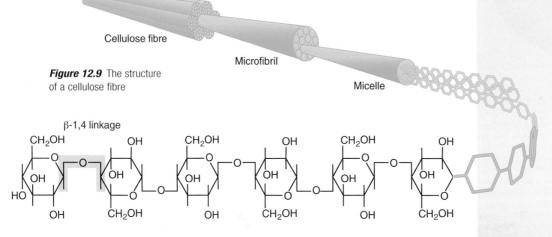

Cellulose fibre

Microfibril

Micelle

Figure 12.9 The structure of a cellulose fibre

The cellulose fibres are then used to build a cell wall. They are not all aligned in the same direction, but 'criss-cross' each other as shown in the micrograph. This arrangement gives both strength and some elasticity.

Why do plant and animal cells store different carbohydrates?

This is related to the different metabolic needs of plants and animals. Enzymes in plants polymerise α-glucose to either amylose or amylopectin, which together make starch. In animals, different enzymes (resulting from different DNA) polymerise α-glucose into glycogen.

(×14 000)

Electron micrograph showing the cellulose fibres in a plant cell wall

Amylose is a spiral molecule with no side branches. Amylopectin has side branches. So what is the advantage of having two storage products? Amylose is a more compact molecule than amylopectin. Therefore, more can be packed into a starch grain. However, it only has two ends! This means that when hydrolysed to α-glucose, the process can only take place at the two 'ends', so it is a relatively slow reaction. Amylopectin is branched, so hydrolysis can take place at all the 'ends' and is, therefore, faster. However, the molecule takes up more space. Both amylose and amylopectin are insoluble, which means that they have no osmotic effects on surrounding cells.

(a) α-1,4 linkage

CH₂OH (repeated)

◀ If a cell 'stored' glucose at a high concentration, the cell contents would have a low water potential and would draw water from the surrounding cells by osmosis.

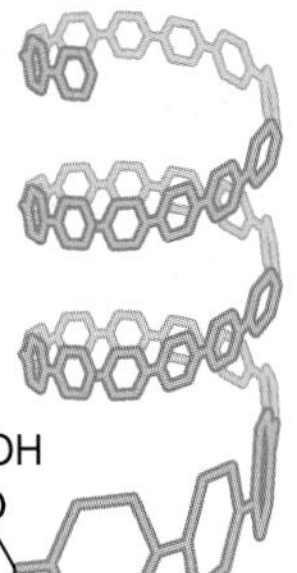

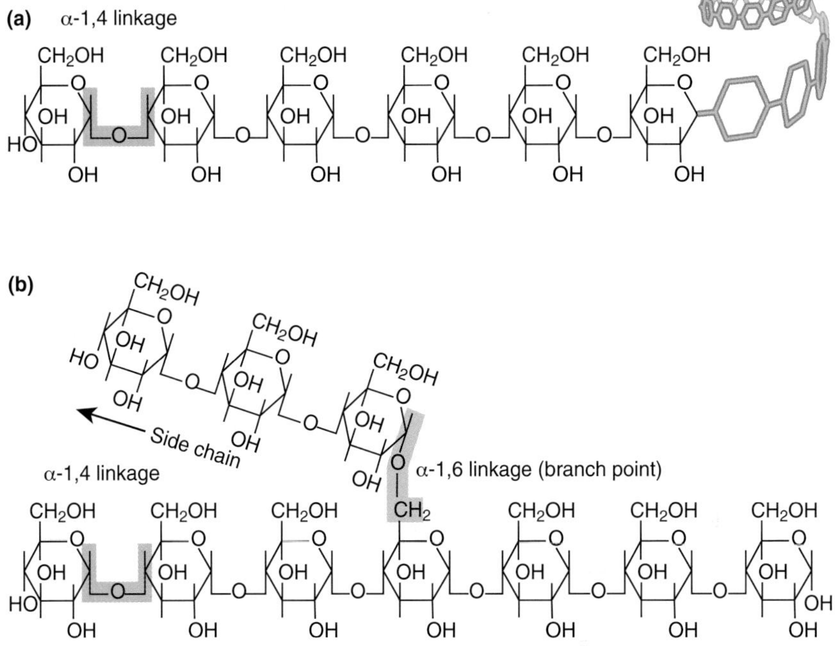

Figure 12.10
(a) Structure of amylose
(b) Structure of amylopectin

(b)

Side chain

α-1,4 linkage   α-1,6 linkage (branch point)

Animals also need to store carbohydrate, but, because of their higher metabolic rate, they need to be able to 'access' the store quickly. This means that their storage product (glycogen) must be capable of being hydrolysed rapidly. Glycogen is a highly branched molecule — much more branched than amylopectin.

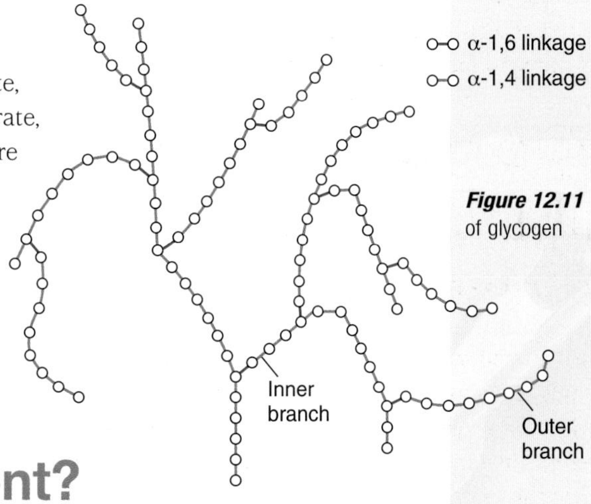

o–o α-1,6 linkage
o–o α-1,4 linkage

Figure 12.11 Structure of glycogen

Inner branch

Outer branch

Why are plant and animal cells different?

Bearing in mind that we humans have 50% of our DNA in common with a banana, the 50% difference in DNA leads to considerable biochemical variation. These biochemical differences between plant and animal cells result in major structural differences. If we consider a palisade mesophyll cell from a leaf, and an epithelial cell from the small intestine, we can see some of these differences, even through an optical microscope (Figure 12.12).

Figure 12.12 (a) The structure of a palisade mesophyll cell and (b) the structure of an epithelial cell from the small intestine, as seen through an optical microscope

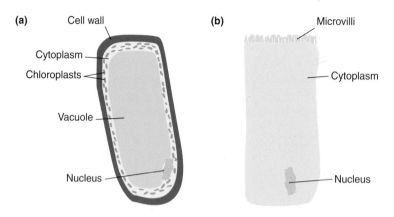

(a) Cell wall
Cytoplasm
Chloroplasts
Vacuole
Nucleus

(b) Microvilli
Cytoplasm
Nucleus

The plasma membrane is not labelled in either of the diagrams because it cannot be seen under a light microscope. What appears as the 'edge' of the animal cell is more of a shadow effect than the plasma membrane itself.

Table 12.1 compares the structures seen in 'typical' plant and animal cells though an optical microscope.

Of course, when seen through an electron microscope, there are other differences, but those given in Table 12.1 are the major ones.

Not all plant cells are like a palisade cell. Many do not have chloroplasts. Some lose their cytoplasm and die soon after they are formed. These are adaptations to the particular functions of particular cells.

Table 12.1 Structures seen in 'typical' plant and animal cells through a light microscope

Structure	Plant cell	Animal cell
Cell wall	Present	Absent
Nucleus	Present	Present
Chloroplast	Present	Absent
Vacuole	Present	Absent
(Plasma membrane	Present	Present)

Figure 12.13 The structure of plant cells is related to their function

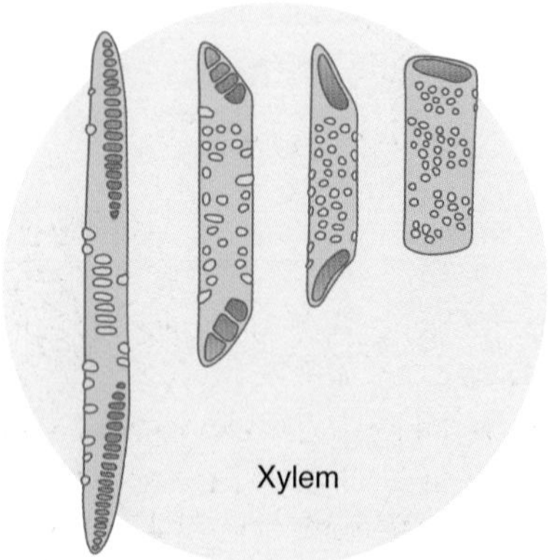

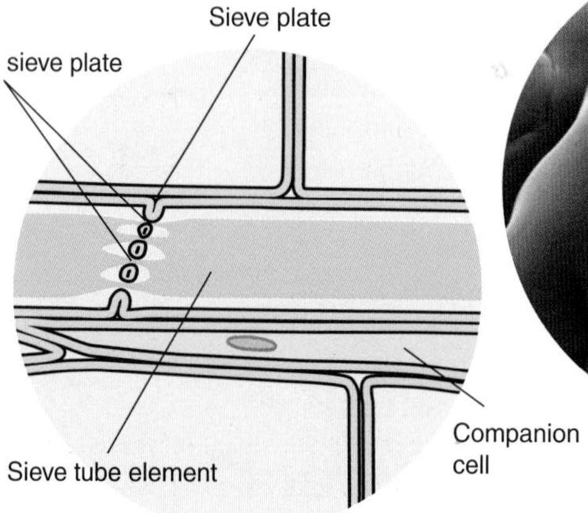

Sieve plate
Pores of sieve plate
Xylem
Sieve tube element
Companion cell
Phloem

(×600)

Dr Jeremy Burgess/SPL

Xylem 'cells' are tubular, dead and completely hollow to allow the efficient transfer of water. These tubular cells join end to end to form continuous water-conducting tubes from root to leaf. (See Chapter 13, pages 231–33.)

Phloem sieve tubes are tubular cells, like xylem. They join end to end with the sieve plates, allowing movement along the tubes. Phloem cells are living cells that move nutrients from sources (where they are formed) to sinks (where there are used or stored).

Guard cells open and close stomata (see Chapter 13, pages 230–31). To do this, they must change shape. This is possible because they have unevenly thickened cell walls that cause the cells to open the stoma when the guard cells take in water, early in the morning.

Why are there different types of human cells?

Surely this can't be due to different DNA, can it? All our cells have the same DNA, don't they? Yes, they do — but all our genes don't 'work' in all our cells. The gene responsible for black hair does not work in cells in the big toe! So, different DNA works in different cells. How does it 'know' where and when to work?

Following fertilisation, the zygote divides repeatedly by mitosis to form a hollow ball of cells known as the blastocyst, with a group of cells at one end called the inner cell mass. These cells eventually give rise to nearly all the different adult cells. Through a complex signalling system, the cells in different positions in the embryo develop in different ways into different tissues (Figure 12.14).

◀ This process is called **cellular differentiation**.

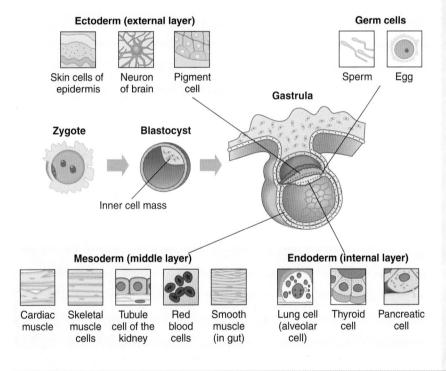

Figure 12.14 Different regions of the embryo give rise to different tissues

Box 12.1 The stem cell debate

Stem cells are cells that retain the ability to renew themselves through mitosis. However, there are different types of stem cell. A stem cell taken from muscle tissue can divide by mitosis to renew itself and, at the same time, form new muscle cells. A stem cell taken from bone marrow can divide to renew itself and, at the same time, form the many types of blood cell. Both these types of stem cell are called **adult stem cells**. Some adult stem cells can also form a limited number of specialised cells different from those in the tissue where the stem cell is found. A stem cell taken from the inner cell mass of a blastocyst can divide to renew itself and can also form any type of cell in the body. A stem cell taken from yet an earlier stage can also do this, but has the ability to become an entire organism — in effect it behaves like a zygote. These last two types are called **embryonic stem cells**.

Because of their ability to develop into any tissue, embryonic stem cells have the potential to be used in the treatment of degenerative diseases. Injection of such stem cells could, hopefully, lead to the regeneration of healthy tissue. Adult stem cells are more limited in their potential and each type has only limited uses.

However, the only way to obtain human embryonic stem cells is from a human embryo. This results in the destruction of the embryo and, to some people, this represents the potential loss of a human life. Some people define the moment of fertilisation as the beginning of a life; others believe that the potential for life does not begin until the embryo has implanted in the uterine lining. This affects people's views on the status of an embryo.

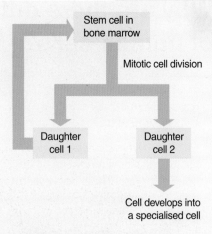

Much of the research into embryonic stem cells has been carried out with mice because work on human embryos has been banned in many countries. However, limited research on human embryonic stem cells has been undertaken with promising results. In the UK, such research is regulated by the Human Fertilisation and Embryology Authority (HFEA). The establishment of stem-cell banks is reducing the need to use 'new' embryos.

In November 2007, researchers reported being able to 'reprogramme' adult stem cells genetically into the embryonic state. This is an exciting discovery. If it is confirmed, it would potentially enable embryonic stem cell therapy without the need to destroy human embryos.

What is a tissue?

Muscle, nerve and bone are examples of tissues. Each tissue contains similar cells, which carry out a specific function. So, we can define a tissue in the following way:

> A tissue is a group of similar cells that all perform the same function.

Some examples of tissues are described below.

Epithelia

An **epithelium** is a layer of cells that covers a body part of an animal. Usually it consists of a single layer of cells sitting on a thin, glycoprotein 'basement membrane'. However, some epithelia are several cells thick; these are called **stratified epithelia**. If the epithelium lines an organ, it may be referred to as an **endothelium**. Unlike many other types of cell, most epithelial cells retain the ability to divide. This is important because cells that line or cover structures are continuously worn away and have to be replaced.

◀ A layer of cells covering a plant organ is called an epidermis.

Cells of **squamous epithelium** are extremely thin. In the lungs, they form the walls of the alveoli and aid gas exchange by contributing to the short diffusion pathway. In the arteries, they provide a smooth inner lining to the walls, reducing resistance to the flow of blood. The **columnar epithelial** cells lining the small intestine have **microvilli** to increase the surface area for uptake of nutrients.

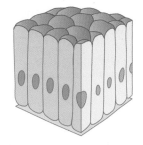

(a)

(b)

Figure 12.15
(a) Squamous epithelial cells
(b) Columnar epithelial cells

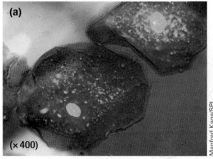

(a)

(×400)

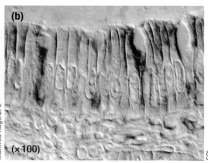

(b)

(×100)

Manfred Kage/SPL

SPL

(a) Squamous epithelial cells
from the mouth
(b) Columnar epithelial cells
in the small intestine

Connective tissue

Connective tissue consists of relatively few specialised cells in a non-cellular matrix that contains protein fibres. Bone and cartilage are typical examples of connective tissue. The non-cellular matrix of bone contains calcium salts that confer hardness. Different types of cartilage contain different proportions of protein fibres. This results in varying degrees of elasticity or rigidity.

Blood

Blood does not conform to the definition that a tissue is 'a group of similar cells all performing the same function'. Blood consists of several types of cell transported by a liquid — the plasma. Blood can be thought of as a kind of connective tissue in which the plasma is the non-cellular matrix and the blood cells are the specialised cells. Blood has more than one function, so there is more than one type of cell.

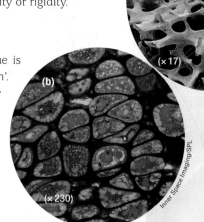

(a)

(×17)

Steve Gschmeissner/SPL

(b)

(×230)

Inner Space Imaging/SPL

(a) Bone and
(b) cartilage are
examples of
connective tissue

Tissues and organs

Organs are structures within an organism that are made of several types of tissue. Each tissue performs its own function and is essential to the overall functioning of the organ.

The heart is the organ that pumps blood around the body. To do this, it contains:
- **cardiac muscle** — to provide the force to pump the blood as the muscle contracts

- blood in blood vessels in the wall of the heart — to carry oxygen to the cardiac muscle cells so that aerobic respiration can take place to release the energy needed for their contraction
- specialised conducting tissue (**Purkyne tissue**) to carry impulses to the cardiac muscle and cause contraction

In other words, the cardiac muscle contracts, the blood transporting oxygen and glucose allows the release of energy for the contractions and the Purkyne tissue 'tells' it when to contract.

Arteries and veins are also organs because they contain several types of tissue. However, capillaries contain only endothelial tissue and so cannot be classified as organs.

A leaf is an organ in photosynthetic plants. It contains several tissues:
- upper and lower **epidermis**, to protect the leaf from damage, infection and dehydration; the lower epidermis also allows gas exchange
- **palisade mesophyll** — the main photosynthetic layer, with columnar cells (allowing tight packing), each containing many chloroplasts to ensure maximum light absorption
- **spongy mesophyll**, to allow diffusion of gases in both directions between the atmosphere (via the stomata) and the palisade layer
- veins containing:
  - **xylem** to transport water to the leaf
  - **phloem** to transport organic substances to and from the leaf

◄ A muscle (such as the biceps) is an organ.
It contains skeletal muscle tissue, together with arteries and veins (each made from epithelia, smooth muscle and connective tissue), blood and nervous tissue.
A nerve (such as the vagus nerve) contains mainly nervous tissue, but also contains blood vessels, blood and connective tissue.

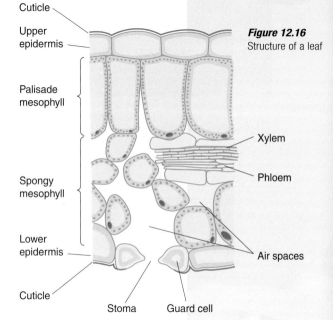

Figure 12.16
Structure of a leaf

Cuticle
Upper epidermis
Palisade mesophyll
Spongy mesophyll
Lower epidermis
Cuticle
Stoma
Guard cell
Xylem
Phloem
Air spaces

Organ systems

Major body processes are not usually performed by single organs but by groups of organs working together. If we consider the circulation of blood, the heart provides the force to move blood through a system of blood vessels, which carry the blood to all parts of the body. Together, the heart, arteries, veins and capillaries form the circulatory system. The breathing system comprises the lungs, trachea, larynx and nasal cavity, as well as the diaphragm and intercostal muscles that make breathing movements possible. Other organ systems include:
- the digestive system
- the nervous system
- the musculoskeletal system
- the reproductive system
- the excretory system

◄ See if you can work out which organs make up each system.

Although each organ system is responsible for a major body process, systems do not work in isolation. The circulatory system transports substances between the other systems; the nervous system controls the functioning of the other systems.

Summary

Haemoglobin

- Haemoglobin is a protein with a quaternary structure; each polypeptide chain contains a haem group that can bind with oxygen
- Haemoglobin loads (binds with/associates with) oxygen when the partial pressure of oxygen is high and unloads (releases/dissociates) oxygen when the partial pressure of oxygen is low.
- An increase in the partial pressure of carbon dioxide decreases affinity of haemoglobin for oxygen (at any partial pressure of oxygen) making it unload extra oxygen; this is the Bohr effect.
- The haemoglobin of a fetus and of animals living where partial pressures of oxygen are low has a higher affinity for oxygen; the oxygen dissociation curve is shifted to the left.

Plant and animal cells

- Cellulose is a polymer of β-glucose molecules; it forms unbranched chains which can be organised into cellulose fibres that make up plant cell walls
- When an animal cell is viewed through a light microscope, the only structures visible are:
  - the cytoplasm
  - the nucleus
- In addition to the structures seen in animal cells, a palisade mesophyll cell has:
  - a cellulose cell wall
  - a vacuole
  - chloroplasts
- Plant and animal cells contain a plasma membrane, although it is not visible through a light microscope

Cells, tissues and organs

- Different human cells result from cellular differentiation. This takes place during the development of the embryo, when the position of a cell in the embryo determines how it will develop.
- A group of similar cells all performing the same function is a tissue.
- Epithelia are tissues that line or cover organs in animals; they usually consist of a single layer of cells.
- Blood is an unusual connective tissue in which several types of cell are suspended in a liquid (the plasma).
- An organ is a structure that consists of several different tissues, each performing different functions that contribute to the overall functioning of the organ.
- An organ system consists of a number of organs that, together, carry out a major bodily process (e.g. transport by the circulatory system).

Questions

Multiple-choice

1 The main advantage of the high degree of branching in a molecule of amylopectin is that:

 A the many 'ends' allow rapid hydrolysis

 B much can be stored in a small space

 C there are no osmotic effects

 D it is insoluble

2 Haemoglobin is:

 A a protein containing four polypeptide chains

 B a protein containing two types of polypeptide chains

 C both A and B

 D neither A nor B

3 Which of the following statements about cellulose is *not* true?

 A it is a polymer of β-glucose

 B the molecule is unbranched

 C the glucose molecules are joined by β-1,6 bonds

 D adjacent molecules form hydrogen bonds with each other

4 Cellular differentiation involves:

 A expression of some genes but not others

 B position of a cell determining how it will develop

 C continued division of the cells

 D all of the above

5 Squamous epithelial cells in the lungs increase the efficiency of exchange of gases by diffusion because:

 A they have microvilli to increase their surface area

 B they rest on a glycoprotein basement membrane

 C they are extremely thin

 D their shape helps to maintain a concentration gradient

6 A tissue is:

 A a group of similar cells carrying out the same function

 B a group of similar cells carrying out different functions

 C a group of different cells carrying out the same function

 D a group of different cells carrying out different functions

7 It is true of the Bohr effect that it:

 A shifts the oxygen dissociation curve to the left and is caused by increased $p\text{CO}_2$

 B shifts the oxygen dissociation curve to the left and is caused by decreased $p\text{CO}_2$

 C shifts the oxygen dissociation curve to the right and is caused by decreased $p\text{CO}_2$

 D shifts the oxygen dissociation curve to the right and is caused by increased $p\text{CO}_2$

8 The aorta contains smooth muscle, elastic tissue and endothelium. It is, therefore:

 A a tissue

 B an organ

C an organ system

D some other structure

9 It is true of organ systems that:

A they consist of several organs working together

B each system carries out a major body process

C the breathing system, the circulatory system and the nervous system are examples

D all of the above

10 When viewed through an optical microscope, palisade mesophyll cells have:

A all the structures in a typical animal cell

B all the structures in a typical animal cell plus a cell wall

C all the structures in a typical animal cell plus a cell wall and a vacuole

D all the structures in a typical animal cell plus a cell wall, a vacuole and chloroplasts

Examination-style

1 The biceps muscle is an organ. It contains several tissues, including muscle tissue, nervous tissue and blood (in blood vessels).

(a) What is a tissue? *(2 marks)*

(b) Suggest the function of each of the tissues named in the overall functioning of the biceps. *(3 marks)*

(c) Explain why the biceps muscle is classified as an organ but muscle is a tissue. *(2 marks)*

Total: 7 marks

2 The diagram shows the oxygen dissociation curve for human haemoglobin at two different pHs.

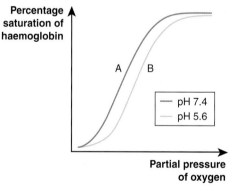

(a) The *y*-axis of the graph is labelled 'percentage saturation of haemoglobin'. Explain what this means. *(1 mark)*

(b) Vigorous exercise could produce an oxygen dissociation curve similar to curve B. Explain how. *(4 marks)*

(c) (i) The properties of fetal haemoglobin differ from those of adult haemoglobin. Describe how. *(2 marks)*

(ii) Explain the advantage of the difference in properties you described in answer to (c) (i). *(3 marks)*

Total: 10 marks

3 A group of people living at high altitudes in South America were found to have higher than usual concentrations of nitrous oxide in their blood plasma. At these altitudes, the partial pressure of oxygen is only 50% of that at sea level.

(a) Scientists investigated the effect of nitrous oxide on the oxygen dissociation curve of haemoglobin. Their results are summarised in the graph below.

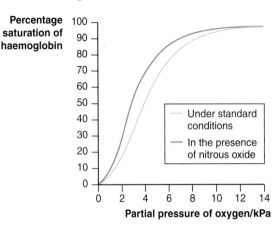

(i) At a partial pressure of oxygen of 8 kPa, there is a difference in the percentage of haemoglobin saturated with oxygen under the two conditions. What is this difference? Show how you arrived at your answer. *(2 marks)*

(ii) Explain how this effect would be an advantage to people living at high altitudes. *(3 marks)*

(b) Nitrous oxide is also known to be a vasodilator; this means it causes blood vessels to dilate (become wider). How could this effect be an advantage at high altitudes? *(3 marks)*

Total: 8 marks

4 Heart attacks are often fatal because cardiac muscle tissue dies and cannot be replaced. Scientists thought that injecting the damaged heart with stem cells might help recovery. They investigated this in mice. Their results are shown in the graphs below.

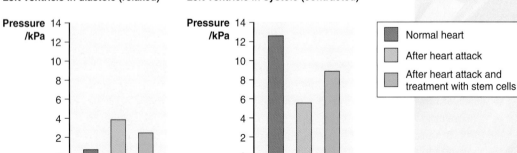

(a) (i) Why is cardiac muscle referred to as a tissue? *(2 marks)*

(ii) Why is dead cardiac muscle tissue not normally replaced? *(2 marks)*

(b) (i) What is the difference between the ventricular pressure in diastole (relaxed) and in systole (contracting) in the normal heart? *(2 marks)*

(ii) The difference in these pressures is altered after a heart attack. Describe how and suggest a reason for the difference. *(3 marks)*

(c) (i) Scientists decided to inject the damaged hearts with stem cells. Suggest why. *(2 marks)*

(ii) The hearts injected with the stem cells showed 68% replacement of dead tissue. Suggest how this explains the results for the hearts injected with stem cells. *(4 marks)*

Total: 15 marks

Chapter 13

Size matters

This chapter covers:
- the relationship between size and surface area
- how surface area-to-volume ratio influences how gases are exchanged
- gas exchange surfaces in a protoctistan, an insect, a fish and a plant
- how surface area-to-volume ratio affects heat gain and heat loss
- movement of water through a plant
- the blood system of a mammal

The first organisms were unicellular (single-celled). They had no need for complex organs or systems. They obtained all the substances they needed by absorption across their plasma membranes by the processes described in Chapter 4. Their metabolic waste was also excreted across their plasma membranes. Over billions of years, the unicellular organisms evolved into complex multicellular organisms. In multicellular organisms, these simple ways of obtaining and removing substances no longer sufficed. Tissues, organs and organ systems evolved.

Why are insects small and how can some trees get so big?

How are size and surface related?

Very small organisms obtain oxygen by simple diffusion through their surfaces. The demand for oxygen is satisfied by the rate at which it can be supplied. However, in bigger organisms, simple diffusion cannot supply enough oxygen and so specialised organs have evolved.

Following on from Fick's law (Chapter 4), key features in determining the rate of diffusion of oxygen (the supply rate) are:

- the surface area of the organism
- the concentration gradient
- the temperature

However, for two different sized organisms, it is the surface area that is most important.

The oxygen supply rate is determined largely by surface area.

The amount of oxygen needed (the demand) is influenced by a number of factors — for example, the temperature, which affects the rate of metabolic reactions. However, at any given temperature, a large organism needs more oxygen than a small organism of the same type.

The demand for oxygen is determined largely by the volume of the organism.

As long as supply can keep pace with demand, diffusion through the body surface as a method of obtaining oxygen remains effective.

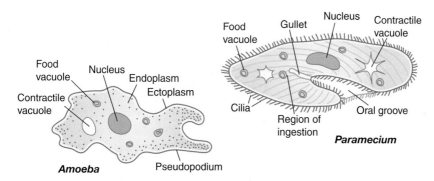

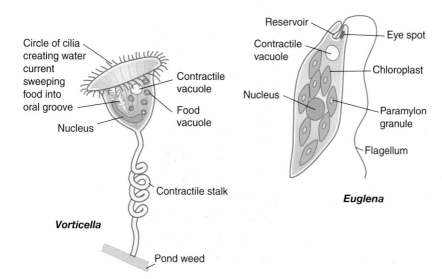

Figure 13.1 These small, single-celled protoctistans obtain oxygen through their plasma membranes by simple diffusion

A measure of the efficiency of this process is the **surface area-to-volume ratio** of an organism. A large ratio indicates that the process is likely to be efficient; a small ratio, that it is not so efficient. We can show mathematically what happens to this ratio as organisms get bigger. For simplicity, imagine that the organism is a cube, and the length of each edge is 1 arbitrary unit (au):

- the area of each face is $1 \times 1 = 1$ au^2
- there are six faces to a cube, so the total surface area is 6 au^2
- the volume is $1 \times 1 \times 1 = 1$ au^3
- the ratio of surface area to volume = 6/1 = 6

Consider a cubic cell in which the length of each edge is 2 au. The total surface area is $6 \times 2 \times 2 = 24$ au^2 and the volume is $2 \times 2 \times 2 = 8$ au^3. The surface area-to-volume ratio is 24/8 = 3 — lower than that of the smaller cube.

If you calculate the values for cubes of edge 4 au and 8 au, you will find that the surface area-to-volume ratios are 1.5 and 0.75 respectively.

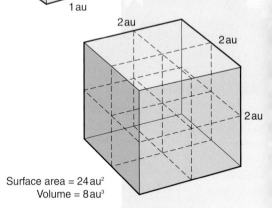

Figure 13.2 The effect of increasing size on the surface area and volume of a cube

1 au   1 au   1 au

Surface area = 6 au^2
Volume = 1 au^3

2 au   2 au   2 au

Surface area = 24 au^2
Volume = 8 au^3

Of course, living organisms are not cubic. However, the principle, that as organisms become larger the ratio of surface area to volume becomes smaller, holds true.

As organisms increase in size, and their volumes increase to a greater extent than their surface areas, the demand for oxygen outstrips the ability to supply it through the body surface. At some point, uptake of oxygen through the body surface becomes too inefficient, and more efficient organs for gaseous exchange are required. An important feature of all these specialised organs is that they have a large surface area. They restore the high surface area-to-volume ratio of the entire organism that is needed for the supply of oxygen to meet the demand. Examples of specialised gaseous exchange organs include:

- lungs — millions of **alveoli** provide the large surface area (see Chapter 6, page 103)
- gills — the many **gill lamellae** provide the large surface area
- leaves — **spongy mesophyll** provides the large surface area

(a)

Corel

(b)

Corel

(a) Whales have lungs to obtain oxygen
(b) Fish have gills to obtain oxygen

e You should be able to distinguish between a *gas-exchange organ* and a *gas-exchange surface*. Lungs are organs specialised for gas exchange, butthe gas-exchange surface is provided by the alveoli. Similarly, gills are gas-exchange organs, but the gas-exchange surface is provided by the gill lamellae.

(c)

Corel

(d)

Arco Images/Alamy

As well as specialised gas-exchange organs, organs specialised for feeding and digestion have evolved, together with organs for reproduction, excretion and movement. Once parts of the body became specialised for different functions, a **transport system** was necessary to move nutrients, gases and excretory products between the specialised body parts. Coordinating systems also evolved to control body functions.

(c) Trees exchange of gases with the environment varies with light intensity
(d) Insects have a tracheal system

How do fish exchange gases with their environment?

Fish have **gills** that extract oxygen dissolved in water. The surface area of each gill is increased by being divided into many **gill filaments**, each of which has many gill lamellae. The walls of the lamellae are very thin, so that the diffusion pathway between water and blood is shortened.

(a) Gill cover in a fish
(b) Gill filaments

Fish draw water into the mouth cavity by creating a negative pressure. They lower the floor of the mouth and water enters the mouth cavity. It is then forced over the gills by closing the mouth and raising the floor of the mouth cavity. As the water passes the gill lamellae, oxygen diffuses from the water into the bloodstream and carbon dioxide diffuses in the opposite direction.

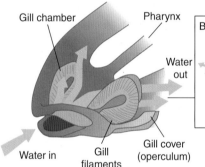

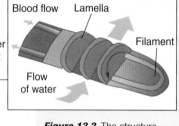

Figure 13.3 The structure of gills

Notice that in the individual lamellae, the blood flows in the opposite direction to the water. This **countercurrent** maintains a concentration gradient for both carbon dioxide and oxygen between water and blood. If water and blood flowed in the same direction, the concentrations would reach equilibrium and gas exchange would cease. So, the countercurrent mechanism improves the efficiency of gas exchange.

Position of gills

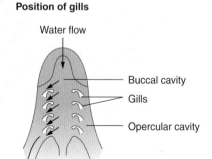

Part of one gill

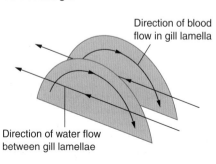

Figure 13.4 Part of one gill

How do insects exchange gases with their environment?

Insects do not have lungs or gills, but a tracheal system that consists of two main **tracheae** running the length of the insect's body. Each trachea opens to the air through several **spiracles** along its length.

(a)

(b)

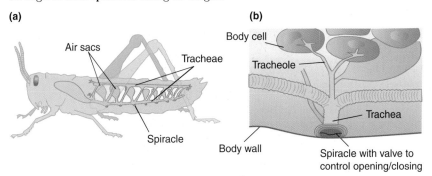

Figure 13.5
(a) The tracheal system of an insect
(b) The relationship between spiracle, trachea and tracheoles

Smaller tubes, **tracheoles**, branch off the tracheae and carry air directly to the cells of the insect's body. The large numbers of tracheoles create a large surface area for gas exchange; the thin walls of the smallest ensure a short diffusion distance. Tracheoles also carry air into the air sacs, which function as temporary stores. When an insect needs to conserve water — for example in very hot conditions — it closes the spiracles and uses oxygen from the air stored in these sacs. This minimises water loss by evaporation.

Movement of gases through the tracheal system happens almost entirely by diffusion in small insects. Oxygen diffuses down its concentration gradient towards the actively respiring cells, and carbon dioxide diffuses in the opposite direction. However, in larger insects there is some ventilation of the system. By opening some spiracles and closing others while dilating and constricting the abdomen, air (rather than oxygen or carbon dioxide) can be moved along the trachea. However, individual gases, rather than air, diffuse to and from cells along the tracheoles.

Box 13.2 Breathing in aquatic insects

Aquatic insects also depend on air and the tracheal system to breathe. For example, mosquito larvae hang suspended upside down from the surface of a pond. They have a breathing tube that connects with their tracheal system and allows them to take in air.

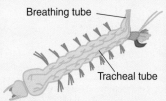

Figure 13.6 A mosquito larva obtains air through a breathing tube

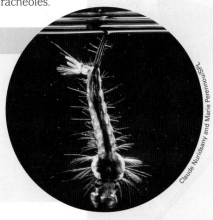

A mosquito larva

Other aquatic insects carry a bubble of air with them as they 'dive'. Many can remain submerged for several minutes by relying on the oxygen in the bubble.

What about gas exchange in plants?

Most gas exchange in plants occurs in the leaves. It is rather more complex than in animals, because plants not only respire; when there is sufficient light, they also photosynthesise.

Respiration takes place all the time:

$$C_6H_{12}O_6 + 6O_2 \rightarrow 6CO_2 + 6H_2O + \text{energy released}$$

Photosynthesis takes place when it is light:

$$6CO_2 + 6H_2O + \text{light energy} \rightarrow C_6H_{12}O_6 + 6O_2$$

So, plants take in carbon dioxide and release oxygen during the day, but take in oxygen and release carbon dioxide at night.

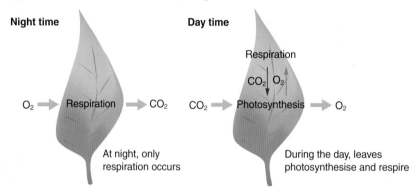

However, exchange takes place in the same way for both gases. They move down their concentration gradients in and out of the leaf through **stomata**.

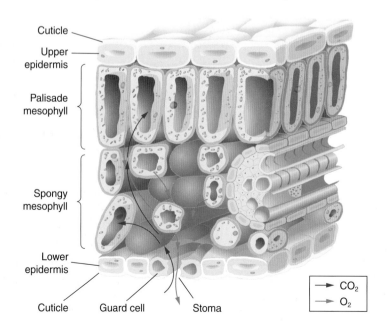

Figure 13.7 Gas exchange in a leaf during the day

The spongy mesophyll provides a large surface area for gas exchange as well as allowing gases to diffuse easily to the palisade mesophyll.

However, water vapour can also pass through open stomata. There is a potential conflict between efficient gas exchange and water loss. The structure of most leaves minimises this conflict in three ways:

- Most stomata are on the lower surface, away from the effects of direct heat from the sun, which would increase evaporation.
- The stomata can be opened and closed.
- The leaves have a waxy cuticle that prevents excessive evaporation of water through the epidermis.

Spongy mesophyll

M. I. Walker/SPL

(× 100)

As we shall see in a later section, **xerophytes** (plants adapted to very dry conditions) have further adaptations.

Table 13.1 summarises the features of gas exchange surfaces.

Table 13.1

	Surface of protoctistan	Fish gill lamellae	Insect tracheoles	Spongy mesophyll	Alveoli
Respiratory medium	Water	Water	Air	Air	Air
Exchange surface	Plasma membrane	Lamellae	Tracheoles	Plasma membranes of spongy cells	Alveolar epithelium
Ventilation	None	Movements of mouth create one-way water current	In large insects, abdomen dilates to decrease pressure and draw air in	None	Movements of diaphragm and ribs create two-way air current
How large surface area-to-volume ratio is created	Small volume of individual cell	Large area of lamellae	Large area of tracheoles	Large area of cell surfaces	Large area of alveoli
Concentration gradient maintenance	Use of oxygen in cell	Countercurrent system in lamellae	Use of oxygen in cells of body	Use of carbon dioxide by mesophyll cells	Ventilation and circulation
Thin exchange surface/short diffusion pathway	Plasma membrane is thin	Only a single layer of cells between blood and water	Thin wall of tracheoles	Only cell wall and plasma membrane separate cells and air spaces	Extremely thin epithelium — capillaries close to alveoli

How is water transported through a plant?

In plants, there is no active pump in the sense of a physical structure such as a heart. However there are mechanisms that physically move water from root to leaf faster than diffusion would allow. There is therefore a **mass-flow** system in plants. The last part of this, the loss of water from leaves, is called **transpiration**.

Water moves through a plant in the following ways:
- it moves from one living cell to another (across the roots and leaf) down a water potential gradient, by **osmosis**
- it moves through the xylem from root to leaf because of a combination of **physical forces**:
  - **root pressure** (a physical push)
  - **capillarity** (a process by which water 'creeps' up very narrow tubes)
  - **tension** (negative pressure) due to evaporation of water from the leaves (a pull)
- it **diffuses** down a water potential gradient from the air spaces of the spongy mesophyll of the leaf, through open stomata and into the atmosphere

How does water enter and move across the root?

Water enters the root epidermal cells, particularly the root hair cells, by osmosis down a water potential gradient. A water potential gradient exists across the root: the epidermal cells have a higher (less negative) water potential than cells in the centre of the root. Therefore, water moves towards the centre of the root where the xylem is found.

◀ The root hair cells increase the surface area available for absorption of water.

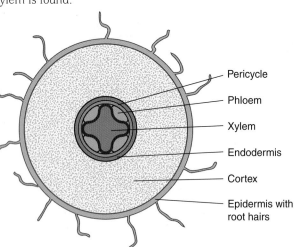

Figure 13.8 Transverse section of a root

Pericycle

Phloem

Xylem

Endodermis

Cortex

Epidermis with root hairs

There are two main pathways by which water moves through the root:
- the **symplast** pathway — in this pathway, water moves through the walls, membranes and cytoplasm of the cells
- the **apoplast** pathway — in this pathway, water moves only through the cell walls and intercellular spaces

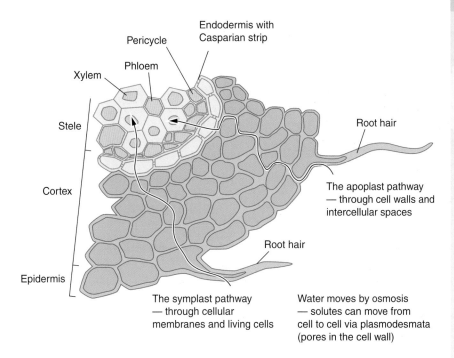

Figure 13.9 Movement of water across a root

The cells of the endodermis have a layer of suberin (a fatty substance) called the **Casparian strip** in their walls. This acts as an apoplast block, preventing water and minerals from entering the vascular cylinder by seeping between the cells (the easiest route). The Casparian strip makes water and minerals pass through the cell membrane.

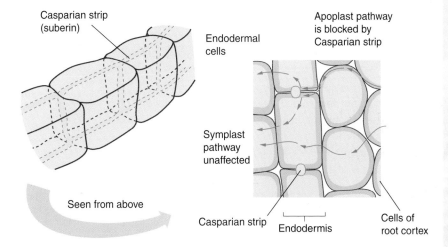

Figure 13.10 Casparian strip

Cells surrounding the xylem elements in the root secrete ions into the xylem, reducing their water potential. Water then follows by osmosis. As more and more water enters the xylem in the centre of the root, it creates a pressure, which forces the water up the xylem. This root pressure is one of the physical forces responsible for moving water up the xylem in roots and stems.

How is water lost from leaves?

A water potential gradient exists from the xylem in the leaf ($\psi = -0.5$ MPa) to the leaf cells ($\psi = -1.5$ MPa) to the air spaces ($\psi = -10$ MPa) and finally to the atmosphere ($\psi = -13$ to -120 MPa). When the guard cells open the stomata, water moves down this water potential gradient.

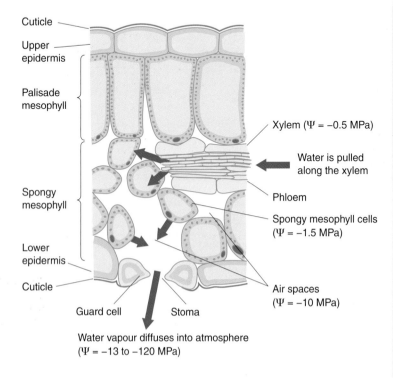

Cuticle

Upper epidermis

Palisade mesophyll

Spongy mesophyll

Lower epidermis

Cuticle

Guard cell

Stoma

Xylem ($\Psi = -0.5$ MPa)

Water is pulled along the xylem

Phloem

Spongy mesophyll cells ($\Psi = -1.5$ MPa)

Air spaces ($\Psi = -10$ MPa)

Water vapour diffuses into atmosphere ($\Psi = -13$ to -120 MPa)

Figure 13.11 How water moves through leaves

(1) When the stomata are open, water vapour diffuses down the water potential (concentration) gradient from the air spaces into the atmosphere.

(2) More water evaporates from the surface of the spongy mesophyll cells into the air spaces, making the water potential of these cells more negative.

(3) Water passes, by osmosis, either from a neighbouring mesophyll cell or from a xylem vessel.

(4) The loss of water from the xylem creates a negative pressure or **tension** in the xylem and the water is pulled along.

All of these processes are dependent on the stomata being open. Transpiration ceases when the stomata are closed (except for a little **cuticular transpiration**, in which some water is lost through the waxy cuticle).

Stomata are opened and closed in the following way. Light activates a pump that moves potassium ions into guard cells from neighbouring epidermal cells. This increase in concentration of potassium ions in the guard cells decreases (makes more negative) their water potential and so they take in water by osmosis. This causes the cells to swell, and because of the alignment of the cellulose fibrils in their cell walls, they become more curved in shape and open the stoma.

In darkness   **Illuminated**

Figure 13.12 How stomata open and close

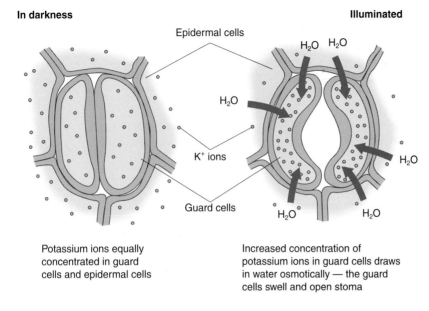

Potassium ions equally concentrated in guard cells and epidermal cells

Increased concentration of potassium ions in guard cells draws in water osmotically — the guard cells swell and open stoma

Figure 13.12 How stomata open and close

How is water moved from root to leaf?

Water moves up the stems in the xylem vessels, which form continuous narrow tubes from roots to leaves.

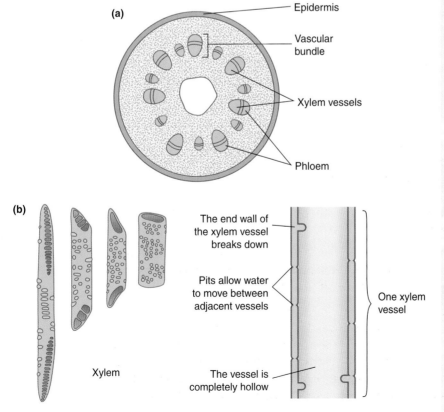

Figure 13.13
(a) Transverse section of a young plant stem
(b) Structure of xylem

Water potential gradients move water from the soil to the xylem in the roots and from the xylem in the leaves to the atmosphere. However, there is no water potential gradient through the xylem. From the roots to the leaves, the water potential in the xylem is almost unchanged. Water moves through the xylem because of other forces.

Possible explanations for the movement of water must account for water being pushed/pulled to heights of more than 30 metres.

Root pressure

As water enters the xylem in the roots by osmosis, it pushes water already present upwards.

Table 13.2

Evidence in favour	Evidence against
Guttation — droplets of liquid water are sometimes seen at the edges of the leaves of small plants (this must be being forced out — normally only water vapour is lost)	Experimental measures of the force due to root pressure suggest that it is sufficient only to force water to a height of 4–5 metres up a stem
Water is pushed out of the cut ends of stems if the root system is still intact	

Root pressure alone is therefore insufficient to explain transpiration in large trees.

Capillarity

The continuous xylem vessels act like narrow capillary tubes. Water molecules adhere to the surfaces of the tubes and creep up the tube, pulling more water molecules with them.

Table 13.3

Evidence in favour	Evidence against
Water does rise several metres up very narrow capillary tubes	Water can only move 1–2 metres up a tube that has the same diameter as a xylem vessel

Capillarity alone is therefore insufficient to explain transpiration in large trees.

The cohesion–tension theory driven by evaporation from the leaves

Evaporation from the spongy mesophyll cells and diffusion into the atmosphere draws water from the xylem in leaves by osmosis. This creates a negative pressure or tension on the water, pulling it upwards. Cohesive forces between water molecules in the xylem are strong enough to resist the tension. A column of water is pulled upwards through the xylem vessels.

The cohesion–tension theory, driven by evaporation from the leaves, is the currently accepted explanation of transpiration.

Evidence in favour	Evidence against
When stomata open in the morning, water movement begins in the leaves, then in the stems	There is no significant evidence against this theory
The tension measured in water in the xylem vessels is great enough to lift a column of water well over 30 metres	
The cohesive forces are strong enough to resist the tension (so the water can move as a continuous column without 'breaking')	

Table 13.4

What factors affect the rate of transpiration?

Any factor that increases loss of water vapour by diffusion will increase evaporation and therefore increase the rate of transpiration. These factors can be grouped into two categories:

- those that affect the water potential gradient between the air spaces in the spongy mesophyll and the atmosphere
- those that affect the total stomatal aperture (effectively, this represents the surface area available for diffusion)

Factors affecting the water potential gradient include:

- atmospheric humidity. A high concentration of water vapour in the atmosphere will increase the water potential of the atmosphere (make it less negative). This will reduce the water potential gradient (concentration gradient of water vapour) between air spaces in the leaf and the atmosphere. The rate of transpiration is reduced.
- atmospheric temperature. When temperature increases, the water vapour molecules have more kinetic energy; they move faster away from the stomata as they escape. This decreases the water potential of the atmosphere (makes it more negative) and so increases the water potential gradient (concentration gradient of water vapour) between air spaces and the atmosphere. The rate of transpiration increases.
- wind. Wind moves water vapour molecules away from the stomata as they escape. This decreases the water potential of the atmosphere and increases the water potential gradient. The rate of transpiration increases.

Factors affecting total stomatal aperture include:

- light intensity
- concentration of carbon dioxide

How can we measure the rate of transpiration?

The apparatus we use is called a **potometer**.

◀ The name of this apparatus is derived from the Greek verb *poton*, to drink

There are two basic types of potometer:

- the bubble potometer, which measures the rate of water uptake by a plant by timing how quickly a bubble in a column of water moves a certain distance along capillary tubing of known diameter
- the mass potometer, which measures the water lost from a plant by measuring the change in mass over a period of time

The bubble potometer

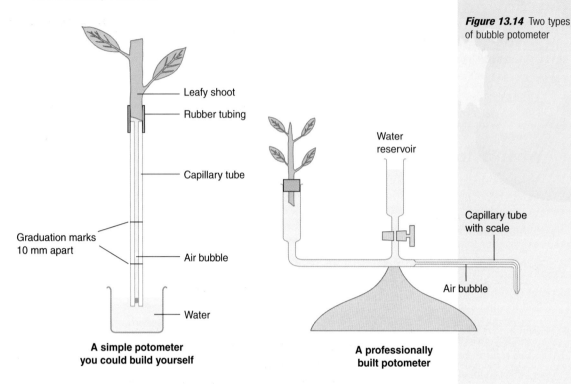

Figure 13.14 Two types of bubble potometer

Leafy shoot

Rubber tubing

Capillary tube

Graduation marks
10 mm apart

Air bubble

Water

**A simple potometer
you could build yourself**

Water
reservoir

Capillary tube
with scale

Air bubble

**A professionally
built potometer**

In both the designs shown in Figure 13.14, the rate of transpiration is measured by timing how long it takes for a bubble of air to move a set distance in mm.

The volume of water taken in can then be calculated (if the diameter of the capillary tubing is known) from the formula

volume $= \pi r^2 l$

where r = radius of capillary tubing ($^1/_2$ diameter) and l = distance moved by bubble

Since the time for this volume of water to be taken in is known, the rate of water uptake per hour can be calculated.

The leafy shoot should be placed in the apparatus under water so that no unwanted air bubbles are introduced. The first design (on the left of Figure 13.14) is easier to assemble and cheaper, but repeat readings are difficult to obtain as the apparatus must be re-assembled each time. The second apparatus is more costly and also slightly more difficult to assemble, but does allow repeat readings to be taken. After each reading, more water can be run into the apparatus from the reservoir, pushing the air bubble back to the end of the capillary tube, ready for another reading to be taken.

Both versions measure water uptake, which is assumed to be directly related to water loss by transpiration.

The mass potometer

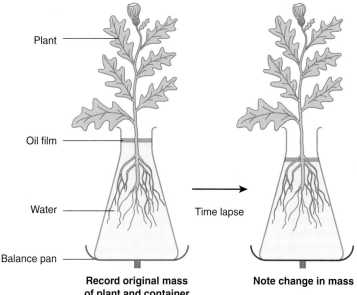

Figure 13.15 The mass potometer

Plant

Oil film

Water   Time lapse

Balance pan

Record original mass   **Note change in mass**
of plant and container

This apparatus does actually measure the amount of water lost, rather than the amount taken up. However, its accuracy is limited by the accuracy of the balance used to measure the mass. The assumption in this apparatus is that water loss from the plant accounts for the entire loss in mass, although some loss in mass could be due to losses through the oil film.

What are xerophytes?

Xerophytes are plants that live in arid (very dry) conditions. To survive, they must either absorb as much water as possible when available, lose as little water as possible through transpiration, or both.

Adaptations shown by xerophytes include:
* extensive root systems to absorb as much water as possible, quickly, when it rains
* reduced leaf area to minimise water loss (the spines of cacti are leaves)
* stems that are capable of:
  – photosynthesis (to compensate for the reduced photosynthesis by small leaves)
  – water storage
* sunken stomata, which lose less water because a high concentration of water vapour accumulates in the 'pits' and reduces the water potential gradient
* epidermal hairs, which also maintain a high concentration of water vapour near to the stomata
* a specialised type of photosynthesis (called CAM photosynthesis) in which the plants only absorb carbon dioxide from the atmosphere at night (so minimising stomatal opening in the extreme heat of the day)

◀ Many cacti have roots that are close to the surface of the soil, but spread several metres from the main stem. When rain does fall, this allows them to absorb a great deal of water quickly, before it drains away.

What about transport systems in mammals?

The structure and function of the heart have been covered in Chapter 5 and the circulatory system was outlined. The system is now considered in more detail.

As we know already from Chapter 5, blood is moved through a system of blood vessels by the pumping of the heart. Mammals have a double circulation, in which, in a complete circulation of the body, blood passes through the heart twice. The main components of this system are shown in Figure 13.16.

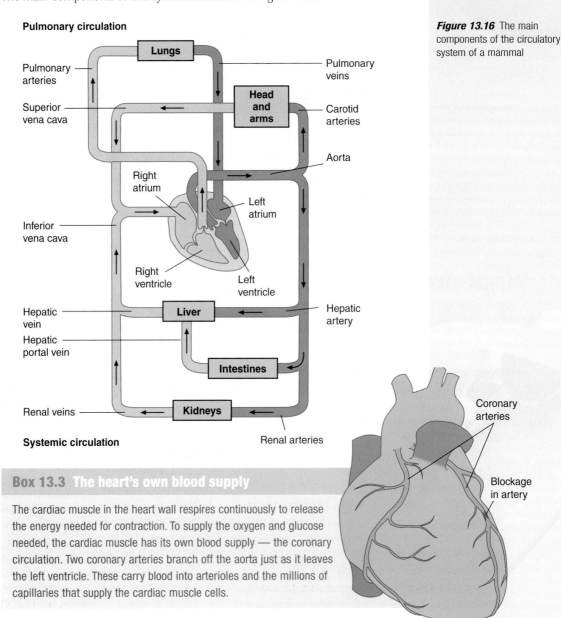

Pulmonary circulation

Lungs
Pulmonary arteries
Superior vena cava
Head and arms
Pulmonary veins
Carotid arteries
Aorta
Right atrium
Left atrium
Inferior vena cava
Right ventricle
Left ventricle
Hepatic vein
Hepatic artery
Hepatic portal vein
Liver
Intestines
Renal veins
Kidneys
Renal arteries

Systemic circulation

Coronary arteries
Blockage in artery

Figure 13.16 The main components of the circulatory system of a mammal

Box 13.3 The heart's own blood supply

The cardiac muscle in the heart wall respires continuously to release the energy needed for contraction. To supply the oxygen and glucose needed, the cardiac muscle has its own blood supply — the coronary circulation. Two coronary arteries branch off the aorta just as it leaves the left ventricle. These carry blood into arterioles and the millions of capillaries that supply the cardiac muscle cells.

What are the blood vessels like?

There are three basic types of blood vessel:

- **Arteries** carry blood under high pressure away from the heart to the organs.
- **Veins** carry blood under low pressure away from the organs towards the heart.
- **Capillaries** carry blood close to every cell within an organ.

The structure of each type of blood vessel is adapted to its function.

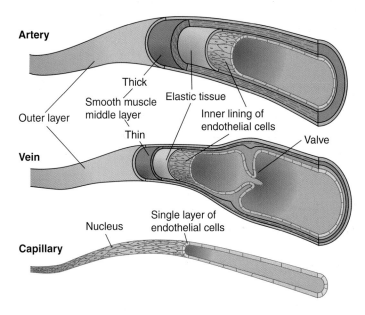

Figure 13.17 Structures of an artery, a vein and a capillary

Arteries

As the ventricles contract, they put the blood under great pressure. As they relax, the pressure drops considerably. As a result, there is a continuous *flow* of blood through the arteries, but the *pressure* fluctuates greatly. We say that the pressure is **pulsatile**. The walls of arteries must not only be able to withstand the high pressures but must be able to compensate for the low pressures. They have a number of tissues in their walls that allow them to do this:

- The outer layer of connective tissue is protective and holds the artery 'open' when the pressure falls.
- The middle layer contains smooth muscle and elastic tissue. This is the thickest layer in the wall of an artery. It allows arterial walls to be stretched when the ventricles contract and cause an increase in blood pressure. It also allows arteries to return to their original diameter when the pressure falls (as the ventricles relax again) — this is called **elastic recoil**. Elastic recoil of the walls of major arteries acts as a secondary pump; as the artery constricts, the force exerted by the wall prevents the pressure of blood from falling too far. The smooth muscle in this layer is innervated. This allows the nervous system to control the diameter of arteries through contraction or relaxation of the smooth muscle in their walls.
- The inner **endothelial layer** creates a smooth surface offering minimal resistance to blood flow.

Arterioles are small arteries that have more muscle in their walls (relative to their size) than arteries. They are also innervated (have nerve endings in their walls). This allows the nervous system to dilate (widen) or constrict (narrow) the arterioles to allow more or less blood through. In this way, blood is redistributed from one organ to another.

Veins

The veins carry blood that has lost most of the pressure created by the ventricles. This is a result of the formation of **tissue fluid** in the capillaries. The walls of capillaries are 'leaky' and as fluid is forced out, the pressure of the remaining fluid drops.

- The wall of a vein contains the same tissues that are present in the wall of an artery, but there are less of them. This is because veins do not have to cope with surges of high pressure or to create elastic recoil.
- The lumen of a vein is larger than that of a similar-sized artery. This allows low-pressure blood to move through the vein with less resistance than there would be if the lumen were narrow.
- Most veins have one-way valves at intervals along their length, to prevent backflow of blood.

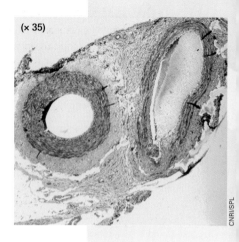

(× 35)

CNRI/SPL

Light micrograph of a section through an artery (left) and a vein (right)

Box 13.4 Moving blood back along veins

As blood flows through capillaries, the pressure falls dramatically because of:

- the formation of tissue fluid
- the increase in total cross-sectional area

When the blood flows back into the veins, the rate of flow recovers but the pressure remains low. So, how *does* blood return from the big toe to the heart?

There are several forces acting on blood in veins:

- **Following blood** — as more blood enters veins from the capillaries, it pushes blood already in the veins further along.
- The **respiratory pump** — on inhaling and exhaling, the pressure in the thorax changes. Inhaling reduces the pressure in the thorax, which means that there is less pressure acting on the veins in that region. This allows the veins to dilate and draw blood into the area. Exhalation increases the pressure and therefore increases the pressure on the veins. The one-way semilunar valves in veins prevent the blood from flowing backwards. It is 'squeezed' into the right atrium.

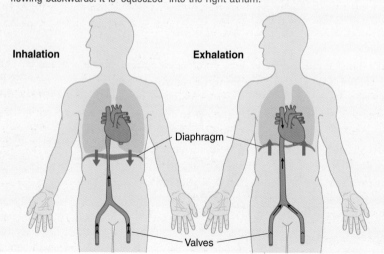

Inhalation   Exhalation

Diaphragm

Valves

The respiratory pump

- The action of **skeletal muscles** — many veins are embedded in skeletal muscles or lie between muscle blocks. In order to maintain body posture, skeletal muscles are continually contracting slightly and then relaxing. This repeatedly puts pressure on veins and then removes that pressure. As the muscles contract, the pressure on the veins forces blood towards the heart. The one-way semilunar valves ensure that there is no backflow of blood.

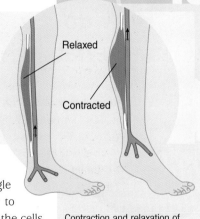

Relaxed

Contracted

Capillaries

Capillaries are microscopic blood vessels. Their walls consist of a single layer of squamous epithelium cells only. Capillaries carry blood near to every cell of an organ. Materials are exchanged between capillaries and the cells of the organ via the tissue fluid.

Contraction and relaxation of the skeletal muscles helps return venous blood to the heart

Table 13.5 compares the structure and function of arteries, arterioles, capillaries and veins.

Table 13.5

Feature	Artery	Arteriole	Capillary	Vein
Cross-section of vessel	Thick wall and narrow lumen	Thinner wall than artery, but relatively more muscle	Microscopic vessels; wall one cell thick	Valves, thin wall, little muscle, large lumen
Blood flow	To an organ, away from the heart	Within an organ, to capillaries in different parts of the organ	Around cells of the organ	Away from an organ, towards the heart
Type of blood	Oxygenated*	Oxygenated*	Blood becomes deoxygenated*	Deoxygenated*
Blood pressure	High and in pulses (pulsatile)	Not quite as high and less pulsatile	Pressure drops through-out capillary network	Low and non-pulsatile
Main functions of vessels	Transport of blood to an organ	Transport in an organ; redistribution of blood	Formation of tissue fluid to allow exchange between blood and cells of an organ	Transport of blood back to the heart
Adaptations to the main function	Large amount of elastic tissue in wall allows stretching due to pulses (surges in blood pressure) and recoil after pulses; endothelium forms a smooth layer to give least resistance	Large amount of smooth muscle under nervous control to allow redistribution of blood; constriction limits blood flow to an area; dilation increases blood flow; constriction of *all* arterioles increases resistance and blood pressure	Small size allows an extensive network close to all cells of an organ; thin, permeable ('leaky') wall allows formation of tissue fluid for exchange with surrounding cells	Large lumen and thin wall offer least resistance to flow as blood is under low pressure; valves prevent backflow of blood

* Except the pulmonary and umbilical blood vessels

How do capillaries exchange materials with cells?

Blood flows close to every cell of the body in the capillary networks in all organs. However, it is **tissue fluid**, not blood, which carries glucose and oxygen to the cells. Tissue fluid is formed from blood in every capillary network. It flows around the cells, bathing them in a fluid that provides a constant environment. The constant pH and temperature of the tissue fluid help to provide optimum conditions for enzyme activity in the cells.

Tissue fluid forms because the capillary walls are permeable to most molecules and the pressure of the blood entering the capillaries is high enough to force materials across the capillary walls. However, **plasma protein** molecules are too large to escape and so are not found in tissue fluid.

As tissue fluid leaves the blood, it carries with it dissolved oxygen and nutrients. These enter the cells from the tissue fluid by one of the transport processes discussed in Chapter 4. Most of the tissue fluid that bathes the cells is returned to the blood in the capillary networks, carrying with it dissolved carbon dioxide and other metabolic waste products. The remainder drains into the **lymphatic system**. The lymphatic vessels carry lymph towards the heart. Lymph is returned to the blood where veins from the head and neck join with those from the arm.

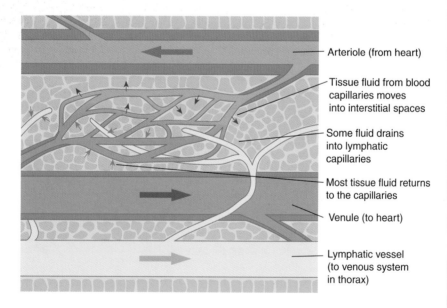

Figure 13.18 The circulation of tissue fluid and the formation of lymph

- Arteriole (from heart)
- Tissue fluid from blood capillaries moves into interstitial spaces
- Some fluid drains into lymphatic capillaries
- Most tissue fluid returns to the capillaries
- Venule (to heart)
- Lymphatic vessel (to venous system in thorax)

How is tissue fluid formed?

Two main factors are involved in the formation of tissue fluid:

- Because of the contraction of the ventricles, there is a high **hydrostatic pressure** in the blood. This acts outwards on the wall of any vessel carrying the blood. If the wall is permeable, liquid is forced out of the vessel.

- The other factor is the **water potential** of blood plasma. The plasma contains many dissolved substances. Therefore, it has quite a low water potential, which tends to draw water into the plasma by osmosis.

At the arterial end of a capillary network, the hydrostatic effect outweighs the effect of the water potential and tissue fluid (all the substances in the plasma except proteins) is forced out of the capillaries. The loss of fluid reduces the hydrostatic pressure of the blood, while the water potential remains more or less unchanged. At the venous end of the capillary network, the effect of the water potential outweighs that of the hydrostatic pressure and water is drawn back into the capillaries by osmosis. Other substances (such as carbon dioxide) diffuse into the blood down concentration gradients.

Plasma proteins that are not forced out of the blood are chiefly responsible for the water potential of the plasma being lower than that of surrounding tissue fluid at the venous end of a capillary network.

Figure 13.19 The forces involved in the formation and reabsorption of tissue fluid

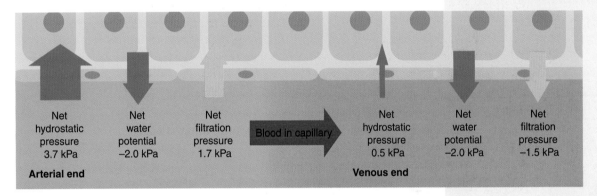

| Net hydrostatic pressure 3.7 kPa | Net water potential −2.0 kPa | Net filtration pressure 1.7 kPa | Blood in capillary | Net hydrostatic pressure 0.5 kPa | Net water potential −2.0 kPa | Net filtration pressure −1.5 kPa |

Arterial end   **Venous end**

Box 13.5 Calculating filtration pressure

Worked example

You could be asked to calculate filtration pressures. Suppose the pressures at the arterial end of a capillary network are:

   hydrostatic pressure = 4.3 kPa
   water potential = −3.3 kPa (remember water potential is always
                                                   a negative value)

Answer

To calculate the net filtration pressure, add the two pressures:

   4.3 + (−3.3) = 1.0 kPa

This is a positive value, so the net filtration pressure is an *outward* force. At the venous end of the capillary network, the pressures might be:

   hydrostatic pressure = 1.6 kPa
   water potential = −3.3 kPa

Adding the two figures gives a value of −1.7 kPa. This is a negative value and so represents an *inward* force.

Summary

Surface area-to-volume ratio

- Small organisms, such as unicellular protoctistans, exchange materials with the environment across their body surfaces; the demand for these materials is influenced by the organism's volume.
- The surface area-to-volume ratio is an indicator of the potential efficiency of exchanging materials across a surface.
- As organisms increase in size, volume increases to a greater extent than surface area, reducing both the surface area-to-volume ratio and the efficiency of exchanging materials and heat across the body surface.

Gas exchange

- Mammals exchange gases with the atmosphere, using lungs. Alveoli create a large surface area for gas exchange (Chapter 6).
- Fish exchange gases in water, using gills. Lamellae in the gills increase the surface area for exchange; a countercurrent flow of water and blood maintains a diffusion gradient.
- Insects exchange gases with the atmosphere using a tracheal system. Oxygen diffuses to individual cells from the main tracheae down ever-smaller tracheoles; water loss in insects is minimised by closing spiracles and using stored air in air sacs.
- Plants exchange gases with the atmosphere through stomata. Gases diffuse through the stomata and into/out of cells in the mesophyll. The spongy mesophyll creates a large surface area for exchange and a diffusion pathway to palisade cells.

Water and plants

- Water loss in plants is minimised by the waxy cuticle, the location of most stomata on the lower leaf surface and the fact that the stomata can be closed.
- Water enters and moves across roots by osmosis; there are two pathways, the apoplast pathway (through cell walls and intercellular spaces) and the symplast pathway (through living cells).
- Water is moved up xylem cells in the stem by tension (negative pressure) created in the leaf. Cohesion of the water molecules allows the water to move as a single column.
- Water diffuses down a water potential gradient from cells in the leaf, through the stomata, to the atmosphere.
- The rate of transpiration is influenced by:
  - temperature — increased temperature increases the rate
  - humidity — increased humidity decreases the rate
  - light intensity — stomata close at low light intensities
  - wind — more water vapour is lost on windy days

- Xerophytes are adapted to dry conditions by having some or all of:
  - **thick** waxy cuticles
  - sunken stomata
  - epidermal hairs surrounding stomata
  - CAM photosynthesis
  - reduced leaf area
  - extensive root systems

Circulatory system in mammals

- Arteries always carry blood away from the heart. An artery has a relatively small lumen and a thick wall that contains a lot of smooth muscle and elastic tissue. This allows the wall to be stretched under pressure and also allows elastic recoil.
- Veins carry blood towards the heart. A vein has a larger lumen than a similar-sized artery and a thinner wall. The larger lumen offers less resistance to the flow of blood under low pressure. Veins have valves to prevent the backflow of blood.
- Capillaries have walls consisting only of squamous epithelium, which allows exchange of materials between the blood and cells.
- Tissue fluid leaves blood at the arterial end of a capillary network because the effect of the hydrostatic pressure of the blood is greater than the effect of its water potential. Water returns to the blood at the venous end of a capillary network because the effect of the blood's water potential is greater than that of its hydrostatic pressure. Other substances diffuse back into the blood at the venous end.

Questions

Multiple-choice

1 As organisms increase in size, the surface area-to-volume ratio:
  A increases
  B remains the same
  C decreases
  D increases to a maximum, then remains steady

2 Squamous epithelial cells in the lungs increase the efficiency of exchange of gases by diffusion because:
  A they have microvilli to increase their surface area
  B they rest on a glycoprotein basement membrane
  C they are extremely thin
  D their shape helps to maintain a concentration gradient

3 The smooth muscle in the wall of an artery allows the wall to:
  A be stretched under high pressure
  B show elastic recoil
  C both A and B
  D neither A nor B

4 Fish exchange oxygen with water at their:

  A gill cover

  B gill lamellae

  C gill arch

  D gill raker

5 Water moves through the xylem in a plant's stem due to:

  A cohesion-tension

  B osmosis

  C diffusion

  D none of these

6 Xerophytic adaptations include:

  A extensive root systems and ordinary photosynthesis

  B limited root systems and ordinary photosynthesis

  C limited root systems and CAM photosynthesis

  D extensive root systems and CAM photosynthesis

7 Constriction of arterioles leading to organ A and dilation of those leading to organ B will:

  A increase the rate of blood flow to A and decrease the rate of flow to B

  B increase the rate of blood flow both organs

  C decrease the rate of blood flow to both organs

  D decrease the rate of blood flow to A and increase the rate of flow to B

8 At the arterial end of a capillary network, the hydrostatic pressure of the blood is 4.5 kPa and its water potential is −1.3 kPa. The net filtration pressure is:

  A 3.2 kPa

  B −3.2 kPa

  C 5.8 kPa

  D −5.8 kPa

9 The rate of transpiration is greatest on a day that is:

  A humid and hot

  B dry and hot

  C dry and cold

  D humid and cold

10 All gas exchange surfaces share:

  A a large surface area and a short diffusion distance

  B a small surface area and a short diffusion distance

  C a small surface area and a long diffusion distance

  D a large surface area and a long diffusion distance

Examination-style

1 The diagram shows how some substances are exchanged between a capillary and some surrounding cells.

  (a) Describe how tissue fluid is formed. *(3 marks)*

  (b) Explain one role of tissue fluid, other than the exchange of substances between blood and cells. *(2 marks)*

  (c) Explain why a capillary cannot be considered to be an organ, whereas an artery can. *(3 marks)*

Total: 8 marks

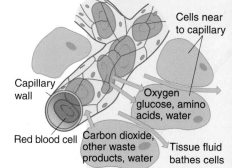

Cells near to capillary

Capillary wall

Oxygen glucose, amino acids, water

Red blood cell

Carbon dioxide, other waste products, water

Tissue fluid bathes cells

2 The diagram shows a transverse section through an artery and a vein.

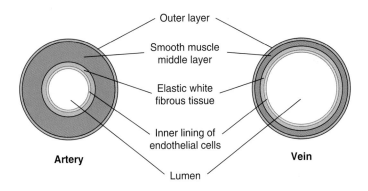

Outer layer
Smooth muscle middle layer
Elastic white fibrous tissue
Inner lining of endothelial cells
Artery
Vein
Lumen

(a) Explain how each of the following features is an adaptation to the functioning of the vessel:
  (i) the amount of smooth muscle and elastic tissue in the wall of an artery *(2 marks)*
  (ii) the size of the lumen of a vein *(2 marks)*
(b) Some aeroplane flights can last over 11 hours. During such flights, because of the long period of inactivity, passengers might experience 'pooling' of the blood in the veins of their legs.
  (i) Explain why the long period of inactivity may lead to pooling of the blood. *(3 marks)*
  (ii) Explain how:
    A normal breathing and
    B walking around the aircraft periodically
       help to return blood in veins to the heart. *(5 marks)*
                                          Total: 12 marks

3 Plants take up water in their roots. The water moves across the root and into the xylem, and then to the leaves.
  (a) Water enters the epidermal cells of a root from the soil and crosses the root cortex to the xylem. Explain how. *(6 marks)*
  (b) Water loss from the leaves is affected by a number of environmental factors, including light intensity.
    (i) List *three* other *environmental* factors that would increase the rate of water loss from a leaf. *(3 marks)*
    (ii) A mass potometer could be used to investigate the effect of humidity on the rate of transpiration. Explain how. *(6 marks)*
                                          Total: 15 marks

4 In an investigation into transpiration, a bubble potometer was used under different conditions;
  Condition 1 Leaves in still air with no vaseline applied
  Condition 2 Leaves in moving air with no vaseline applied
  Condition 3 Leaves in still air with vaseline applied to their lower surfaces
  Condition 4 Leaves in moving air with vaseline applied to their lower surfaces

The results are shown in the graph.

(a) Explain how you would set up the apparatus for this investigation for conditions 3 and 4. *(4 marks)*

(b)(i) For curves A and C, calculate the mean movement per minute for the first 40 minutes. Show your working. *(3 marks)*

(ii) Explain, in terms of water potential, the difference between the two results. *(3 marks)*

(iii) Copy and complete the table to show which condition would be most likely to produce which result. *(3 marks)*

Total: 13 marks

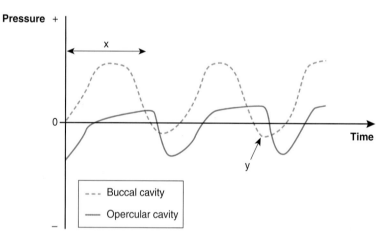

Condition	Curve
1	
2	
3	
4	

5 Fish obtain oxygen from water using gills.

(a) Explain how the gills are adapted to maintain:
  (i) a diffusion gradient between water and blood *(2 marks)*
  (ii) a large surface area *(2 marks)*
  (iii) a short diffusion distance *(2 marks)*

(b) The graph shows how the pressure of water changes in the buccal cavity (mouth) and opercular cavity (region of the gills) over time.

Water enters the mouth at the point marked Y and is pushed over the gills during the region marked X.

Explain how the evidence in the graph supports these statements.

(4 marks)

Total: 10 marks

Chapter 14

Genes, breeding and evolution

This chapter covers:
- the consequences of genetic differences
- how antibiotics are used to treat disease
- how bacteria can become resistant to antibiotics
- vertical and horizontal transmission of antibiotic resistance in bacteria
- how genetic bottlenecks, the founder effect and selective breeding reduce genetic variability in populations
- the effects of in-breeding and out-breeding

The genes that organisms have today are usually those that have given them the best chance of surviving in their current environment. Genes that are disadvantageous are usually not passed on for many generations because the individuals with the disadvantage do not survive to reproduce. Many bacteria today are resistant to antibiotics, because the use of antibiotics is an important part of their environment and so resistance is a useful feature and gives a real survival advantage. If antibiotics weren't used, the feature would have no advantage and few microorganisms would have the genes. Different patterns of breeding in organisms can give the organisms a greater or smaller genetic variability and so affect their chance of survival. Events that reduce genetic variability reduce the survival chances of a species.

What are the consequences of genetic differences?

There may be no significant consequences of genetic differences. For example, some people have DNA that results in them being able to roll their tongues; the DNA of other people results in them not being able to do this. This is not quite the stuff of a serious selective advantage!

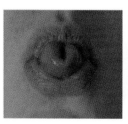 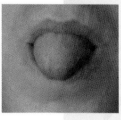

Tongue rolling — some of us can, some of us can't!

However, genetic differences *can* matter. One bacterium could develop resistance to an antibiotic (e.g. one type of penicillin) when the rest of the bacteria in the population do not. This could be a significant development. It could give that bacterium a survival advantage, but *only if* the whole population of bacteria was in an environment where they were *exposed* to the penicillin. If the mutation took place in a remote area where the penicillin was not used, there would be no survival advantage.

Box 14.1 How antibiotics work in treating disease

Antibiotics act against bacteria by disrupting cellular processes such as:

- DNA replication
- protein synthesis
- cell wall synthesis

Some antibiotics kill bacteria — these are **bactericidal** antibiotics. Others do not kill bacteria but stop them from reproducing — these are **bacteriostatic** antibiotics.

The ways in which some different antibiotics act are summarised in the table below.

Mode of action of antibiotic	Example	How the antibiotic works	Bactericidal or bacteriostatic
Disrupts cell wall synthesis	Penicillin	Weakened cell wall cannot resist entry of water by osmosis and cell bursts (osmotic lysis)	Bactericidal
Disrupts DNA replication	Nalidixic acid	Bacteria are not killed, but cell division is halted	Bacteriostatic
Disrupts protein synthesis	Tetracycline	Bacterial cell cannot synthesise enzymes and structural proteins	Bactericidal

Antibiotics that disrupt cell wall synthesis interfere with the synthesis of the peptidoglycan layer in the cell wall. Water enters by osmosis down a water potential gradient. Ordinarily, this entry would be resisted by the cell wall. With only an incomplete cell wall to resist the swelling caused by the entry of the water, the bacterial cell bursts.

Figure 14.1 Action of penicillin on bacteria

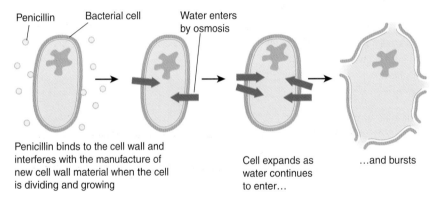

Penicillin   Bacterial cell   Water enters by osmosis

Penicillin binds to the cell wall and interferes with the manufacture of new cell wall material when the cell is dividing and growing

Cell expands as water continues to enter…

…and bursts

Antibiotics are ineffective against viruses. Viruses are acellular and so do not carry out any of the metabolic processes targeted by antibiotics. Viruses enter host cells and, once inside, they are protected by the cells — few antibiotics can enter, because there are no appropriate transport proteins to carry them in.

Box 14.2 The natural selection story

Nearly everyone knows that Charles Darwin first developed the idea of natural selection. However, many of his ideas were not totally new and the way we now interpret the idea of natural selection has 'evolved' from Darwin's original idea.

Some time before Darwin, Jean Baptiste Lamarck had suggested that certain types survive better because they are better adapted to the environment. However, Lamarck suggested that the adaptations were acquired during a lifetime, and that these acquired characteristics were somehow inherited by the offspring. His most widely quoted example is that of giraffes acquiring long necks by stretching to reach foliage in trees — each generation the necks became a little longer because the offspring inherited the stretched necks of the previous generation and then stretched them further!

Darwin as a young man

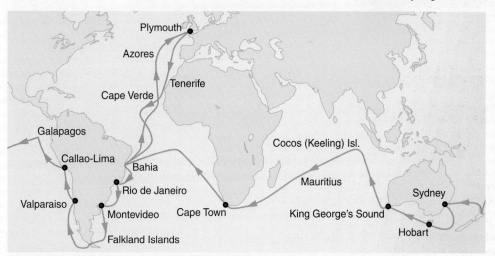

Figure 14.2 The route of the voyage of the *Beagle*

Even though he had no knowledge of genetics, Darwin rejected the idea of acquired characteristics. He made extensive observations while the ship's naturalist on HMS *Beagle*. From these observations he produced two ideas that seemed to apply universally:

- all species tend to produce more offspring than can possibly survive
- there is variation among the offspring

From these observations he deduced that:

- there will be a 'struggle for existence' between members of a species, because they over-reproduce and resources are limited
- because of variation, some members of a species will be better adapted than others to their environment

Combining these two deductions, Darwin proposed:

'Those members of a species which are best adapted to their environment will survive and reproduce in greater numbers than others less well adapted.'

Genes or, more precisely, alleles of genes, determine certain features. Suppose an allele determines a feature that gives an organism an advantage in its environment. As the individuals with the advantage survive and reproduce in greater numbers, the frequency of the advantageous allele in the population will increase.

Mutations introduce new alleles into populations. Any mutation could:

- confer a selective advantage — the frequency of the allele will increase over time
- be neutral in its overall effect — the frequency may increase slowly, remain stable or decrease (the change in frequency will depend on what other genes/alleles are associated with the mutation)
- be disadvantageous — the frequency of the allele will be low and the allele could disappear from the population.

The concept of neutrality was not included in Darwin's original theory. It took an understanding of genetics and mutations before this concept could become included. Also new is the idea that an allele (and therefore the feature it controls) is not adaptive or harmful in itself, but depends on the environment.

Once again, we see how scientific ideas become modified as more and more evidence becomes available.

How do bacteria become resistant to antibiotics?

Bacterial cells are prokaryotic and, like all cells, contain DNA that can mutate. In fact, bacterial cells contain two types of DNA. Most bacterial DNA is organised into a single large, cyclical molecule (as opposed to the linear DNA molecules found in eukaryotic cells); some bacterial DNA is found as *plasmids*. These are small circular molecules of DNA, separate from the main DNA (Chapter 2, page 37).

Antibiotic resistance most often results from mutations in the plasmid DNA.

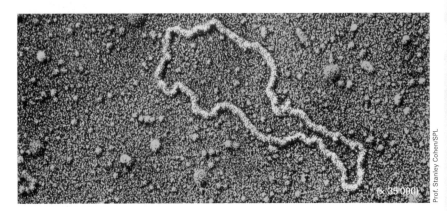

(× 35 000)

Prof. Stanley Cohen/SPL

◀ This is Darwin's now famous theory of **natural selection**. Not a bad achievement for a man who was told at school that he would amount to 'not very much'! However, because he knew little of genetics, Darwin was unable to explain exactly how this happened.

Although opposed by the religious establishment, Darwin's ideas were accepted by most scientists of the day. One of them said 'It is such a brilliantly simple idea that I am annoyed not to have thought of it myself'.

Gregor Mendel's ground-breaking research into genetics would not become widely known and accepted for another 50 years. With our ever-increasing knowledge of genetics, Darwin's theory has been modified.

The large, circular DNA molecule in a bacterial cell

How do bacteria pass on their resistance to antibiotics?

Bacteria can pass on mutations in two ways. One way is when they reproduce and the offspring inherit the mutation. Bacteria reproduce by a process called **binary fission**.

All the DNA in the bacterium replicates prior to reproduction, not just the main DNA molecule. If a plasmid has a mutation that confers resistance to an antibiotic, when the bacterial cell divides each daughter cell will receive some plasmids carrying the gene for resistance. This is sometimes referred to as **vertical transmission**.

> Never refer to binary fission as mitosis. Mitosis is different in several ways, but two key differences are that mitosis:
> - is the division of a *nucleus* (there are no nuclei in bacteria)
> - involves movement of chromosomes on a spindle, neither of which are present in bacteria
>
> The only similarity is that DNA replication takes place prior to cell division.

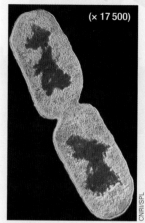

(× 17 500)

CNRI/SPL

Bacteria reproduce by binary fission

Bacterial cell about to divide by binary fission

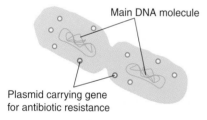

Main DNA molecule

Plasmid carrying gene for antibiotic resistance

Bacterial cells formed from division

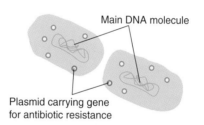

Main DNA molecule

Plasmid carrying gene for antibiotic resistance

Both 'daughter' bacteria are resistant to this antibiotic

Figure 14.3 The gene for antibiotic resistance is passed on when bacteria reproduce by binary fission

◄ Under favourable conditions, some bacteria can divide by binary fission every 20 minutes. In theory, after four hours, there could be over 8000 times as many bacteria as there were initially. After another four hours, there could be 16 000 000 times as many bacteria!

Binary fission is not the only way that bacteria can pass on genes for antibiotic resistance. They can pass plasmids to other bacteria and receive plasmids from them. The process is called **conjugation**.

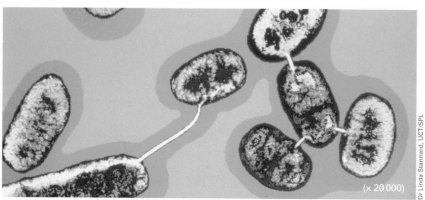

Dr Linda Stannard, UCT/SPL

(× 20 000)

Conjugation in bacteria

Plasmids pass through **conjugation tubes** from one bacterium to another. This is sometimes referred to as **horizontal transmission**.

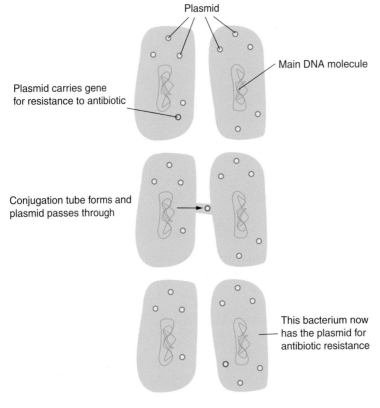

Figure 14.4 How plasmids can pass from one bacterium to another

Plasmid

Main DNA molecule

Plasmid carries gene for resistance to antibiotic

Conjugation tube forms and plasmid passes through

This bacterium now has the plasmid for antibiotic resistance

However, this is not the whole story. It is possible that conjugation is the main mechanism by which some bacteria have become resistant to *several* antibiotics.

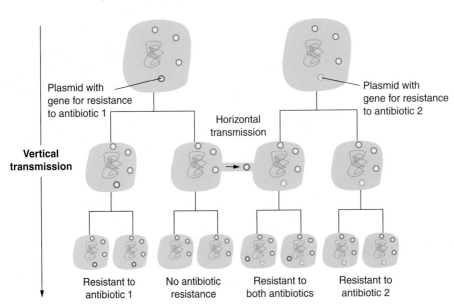

Figure 14.5 The importance of vertical and horizontal transmission in passing on antibiotic resistance in bacteria

Plasmid with gene for resistance to antibiotic 1

Plasmid with gene for resistance to antibiotic 2

Horizontal transmission

Vertical transmission

Resistant to antibiotic 1

No antibiotic resistance

Resistant to both antibiotics

Resistant to antibiotic 2

As we said earlier, the mutation is only advantageous if the antibiotic is being used. If it is being used, this creates a **selection pressure** in favour of those bacteria that have the resistance and against those that do not have it.

More bacteria with resistance survive to reproduce than bacteria without resistance. The frequency of the allele for resistance consequently increases and will increase with each succeeding generation of bacteria, until almost all the population carries the resistant allele. Table 14.1 summarises this.

Selection pressure	Bacterial resistance to antibiotics; repeated use of antibiotic
Variation in the population	Chance mutations in some individuals confer resistance
Which are at an advantage?	Resistant forms
Consequences for phenotype	Resistant forms survive to reproduce in greater numbers — with time, more of the population are resistant
Consequences for allele frequencies	Alleles conferring resistance are passed on in increasing numbers with each generation — frequency increases

Table 14.1

Box 14.3 The MRSA problem

MRSA is one of a generation of 'superbugs' that have arisen in hospitals. MRSA belongs to the species of bacterium called *Staphylococcus aureus* (this accounts for the SA). It is resistant to several antibiotics, including methicillin. MRSA, therefore, stands for methicillin-resistant *Staphylococcus aureus*. It is also resistant to other antibiotics, so some people refer to it as multiply-resistant *Staphylococcus aureus*. Because of its multiple resistance, infections caused by this bacterium are difficult to treat. Therefore, the focus is on prevention of transmission. Programmes to eliminate MRSA from hospitals now concentrate on absolute cleanliness, particularly in wards where patients may have open wounds. It is thought that thousands of people may harbour MRSA without any ill-effects, but that these people could pass on the bacterium to others through skin-to-skin contact. Visitors are now advised to wash their hands thoroughly before entering wards.

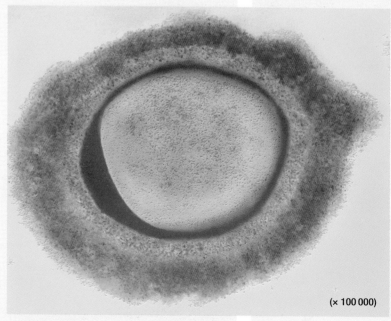

Coloured transmission electron micrograph of a section through MRSA

(× 100 000)

Dr Kari Lounatmaa/SPL

The key here is the use of antibiotics. Some people think that we have been too ready to ask for, and expect to receive, antibiotic treatment for relatively minor infections. If so, the overuse of antibiotics may have provided an increased selection pressure in favour of resistant types of bacteria.

How important is mutation?

Mutation is the only process that produces *new* genetic material. The combined effects of crossing over and random segregation in meiosis and of random fertilisation produce new combinations of existing alleles. As more and more mutations occur within a population, its gene pool (the sum of all the alleles of all the genes in the population) will change. If two populations of the same species are separated for some reason, their gene pools may change in different ways. Over time, the populations may become so different as to be considered two distinct species. Mutation is therefore important in producing the 'raw material' for evolution to occur.

Do any processes reduce the extent of variation?

Any process that reduces a population is likely to reduce genetic variation because it is extremely unlikely that the smaller population remaining will carry a representative sample of all the genes of the previous larger population. Two natural processes that have this effect are:
- genetic bottlenecks
- the founder effect

What is a genetic bottleneck?

A genetic bottleneck is an event in which a population is reduced dramatically and quickly. The survivors cannot carry forward all the alleles from the original gene pool into succeeding generations and so genetic variability is reduced. There is a much smaller gene pool.

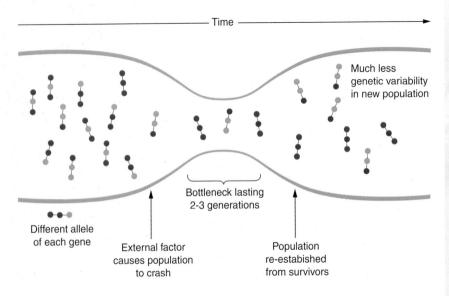

Figure 14.6 Genetic bottlenecks

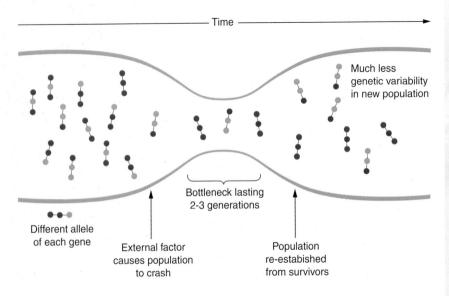Time

Much less genetic variability in new population

Bottleneck lasting 2-3 generations

Different allele of each gene

External factor causes population to crash

Population re-estabished from survivors

Bull northern elephant seal

Gerald C. Kelley/SPL

One striking example of this is to be found in northern elephant seals.

Extensive hunting by humans in the 1890s reduced a population of tens of thousands of northern elephant seal to just 30 individuals. Under protection, the population has grown to 30 000. These are all descended from the 30 survivors of the genetic bottleneck when much of the genetic variability was lost. The southern elephant seals were not hunted to anywhere near the same extent and, as a consequence, show much more genetic variability.

The South African cheetah population suffered a massive genetic bottleneck about 10 000–12 000 years ago. As a result, there is less genetic variability among cheetahs than among other big cats.

The founder effect

The consequences of the founder effect are similar to those of a genetic bottleneck, but the cause is different. The gene pool reduction here occurs when a few individuals from an existing population colonise a new environment. They are the *founders* of a new population.

Because the numbers are low, they are unlikely to be representative of the entire population. Alleles of many genes will not have been 'brought' to the new environment by these founders. So the population that develops from the founders will show less genetic variability than the original population.

The Argentine ant invasion

The first Argentine ants entered the USA in the 1890s in ships carrying coffee from South America. Only a few ants acted as founders of the US population. However, they breed quickly and were soon the most common ant in southern California. Since they were all descended from those first few founders, they were all similar genetically; maybe all were descended from just one queen ant. This had a dramatic effect on their behaviour. In South America, different populations of these ants are territorial and fight ants from neighbouring colonies. In California, because of the genetic similarity, the ants recognise each other as belonging to one 'super-colony'. Millions of ants from different colonies 'team up' to fight insects much larger than themselves. They will even attack bird's nests and kill and feed on the young. They have virtually eliminated one local species of red ant and, with it, the coast horned lizard that feeds largely on them.

This occurred because of the founder effect and the limited genetic variability associated with it. Had many ants from many different ant populations colonised California, they would have fought each other, rather than 'teaming up', and they would not have had such devastating effects on the local ecology.

Argentine ants

Pascal Goetgheluck/SPL

Anthony Mercieca/SPL

Coast horned lizard

How have humans affected the gene pool of other organisms?

Genetic engineering is now the preferred method by which crop plants and stock animals are 'modified' to give increased yields. However, for thousands of years, genetic modification for increased yields was achieved by careful breeding of selected strains. This means only allowing breeding between high-yielding plants or animals, instead of the more random breeding that occurs in wild plants and animals. This more random breeding – **out-breeding** — allows organisms with very different mixes of alleles to breed and maintains genetic variability. When breeding is only permitted between certain types, many of the alleles in the gene pool may be excluded from the population. They will only be passed on if (by chance) they are found in an organism that has desirable alleles for, say, increased milk yield. This so-called **in-breeding** reduces the genetic variability of a population.

Do in-breeding and out-breeding happen naturally?

The short answer is yes, with the same effects as just described. Following a genetic bottleneck or the founding of a new population, the number of potential mates is restricted, so breeding takes place between genetically similar individuals. This is, effectively, in-breeding and it serves to keep the genetic variability low.

In-breeding is sometimes a consequence of religious and cultural beliefs. If, for example, a breakaway religious sect only permits marriage between individuals of the sect, then it is, in effect, ensuring in-breeding and genetic variability within that sect will remain low.

In the case of the cheetah, breeding programmes try to encourage breeding between cheetahs from different populations (out-breeding) in an attempt to increase their genetic variability.

If the programmes are unsuccessful, some people estimate that the South African cheetahs will be extinct in about 20 years.

As variation is the 'raw material' of evolution, out-breeding will ensure variability and give the species the chance to change with environmental conditions and, possibly, evolve into new species. In-breeding limits that variability and makes organisms vulnerable to change. If environmental conditions change significantly, there is a good chance that, because of the lack of variation, all the members of an in-breeding population will be susceptible to the change and none may survive.

South African cheetah — running into trouble?

Summary

- Some genetic differences are of more consequence than others.

Resistance of bacteria to antibiotics

- Mutations can cause bacteria to become resistant to antibiotics.
- Antibiotics act on bacteria by interfering with either:
  - cell wall synthesis (bactericidal) *or*
  - protein synthesis (bactericidal) *or*
  - DNA replication (bacteriostatic)
- Antibiotic resistance usually occurs due to mutations in plasmid DNA.
- Antibiotic resistance is transmitted:
  - vertically (by binary fission)
  - horizontally (by conjugation)
- Horizontal transmission may be responsible for multiple resistance to antibiotics.

Natural selection

- The theory of natural selection proposes that individuals that are best adapted to their environment will survive to reproduce in greater numbers than others; they will pass on their advantageous alleles in greater numbers and so the frequency of those alleles will increase in the population.
- A genetic bottleneck occurs when a population is reduced dramatically in size for at least one generation; many of the alleles are not passed on to future generations, resulting in a reduction in genetic variability.
- The founder effect results in a population with reduced genetic variability because it was founded by a few individuals from an existing population; many of the alleles of the original population are not present in the founders and so are not present in the new population.

Breeding programmes

- Selective breeding focuses on breeding organisms for a particular feature or combination of features; all others are irrelevant to the breeding programme and many alleles are lost from the population.
- Out-breeding encourages genetic variability; in-breeding limits genetic variability.

Questions

Multiple-choice

1 Binary fission is unlike mitosis because:
   A binary fission does not involve a nuclear division
   B binary fission does not involve the movement of chromosomes
   C there is no spindle apparatus in binary fission
   D all of the above

2 The founder effect and genetic bottlenecks are similar because:
   A genetic variability is increased and population size is reduced
   B genetic variability is increased and population size is increased
   C genetic variability is decreased and population size is increased
   D genetic variability is decreased and population size is reduced

3 Transmission of DNA in bacteria occurs:
   A vertically by binary fission and horizontally by conjugation
   B vertically by conjugation and horizontally by binary fission
   C vertically by both binary fission and conjugation
   D horizontally by both binary fission and conjugation

4 Mutation is important in evolution because:
   A it produces new combinations of alleles
   B it produces new DNA
   C it produces new species
   D it produces new populations

5 It is difficult to treat MRSA because it is resistant to:
   A all known antibiotics
   B methicillin
   C penicillin
   D several antibiotics

6 Penicillin kills bacteria by:
   A interfering with DNA replication
   B interfering with protein synthesis
   C interfering with cell wall synthesis
   D it does not kill bacteria; it is a bacteriostatic antibiotic

7 Selective breeding affects genetic variability because:
   A only the alleles affecting high yield are considered
   B other alleles are passed on only if they happen to be present in the high yielding individuals chosen to breed from
   C both A and B
   D neither A nor B

8 Following a genetic bottleneck, genetic variability may increase again if:
   A some in-breeding is possible
   B the mutation rate is abnormally low
   C some out-breeding is possible
   D some of the individuals found a new population

9 Plasmids are:
   A small circular sections of viral DNA
   B small linear sections of bacterial DNA
   C small linear sections of viral DNA
   D small circular sections of bacterial DNA

10 The current theory of natural selection differs from Darwin's theory in that it:
   A recognises that some mutations and some features can be neutral
   B takes genetic explanations into account
   C suggests that it is the combination of genetics and environment that produce, or do not produce, a selective advantage
   D all of the above

Examination-style

1 The graph below shows the effect of a genetic bottleneck on the numbers and genetic variability of a population.

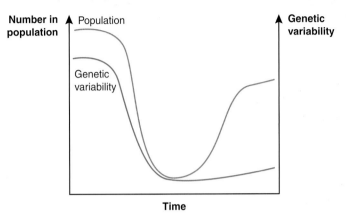

(a) Describe *two* pieces of evidence in the graph that show that a genetic bottleneck took place. *(2 marks)*

(b) Explain *two* possible reasons for the slight increase in genetic variability after the genetic bottleneck. *(4 marks)*

(c) Explain why the population after the genetic bottleneck is more vulnerable to environmental change than the population before the bottleneck. *(4 marks)*

Total: 10 marks

2 The table shows the extent of heterozygosity of three types of DNA in three types of 'big cat': cheetah, puma and lion. Heterozygosity is a measure of the variability of that type of DNA. A large value indicates more genetic variability than a small one.

Organism	Heterozygosity of type A DNA	Heterozygosity of type B DNA	Heterozygosity of type C DNA
Cheetah	0.0072	0.435	0.39
Puma	0.0430	0.579	0.61
Lion	0.0370	0.481	0.66

(a) In which type of DNA was the genetic variability of the cheetah most similar to that of the other two big cats? Use data from the table to justify your answer. *(4 marks)*

(b) The lower genetic variability of the cheetah is thought to be due to a genetic bottleneck that happened 10 000 years ago.

(i) Explain how a 'genetic bottleneck' takes place. *(3 marks)*

(ii) In the captive breeding programme for cheetahs, much emphasis is placed on breeding animals from different populations. Suggest why. *(3 marks)*

Total: 10 marks

3 The diagram below shows two processes that transfer DNA in bacteria.

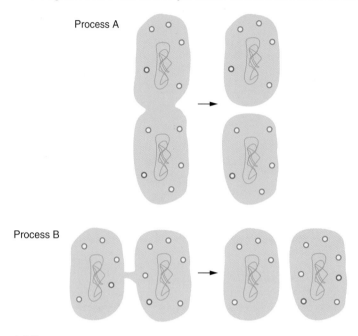

Process A

Process B

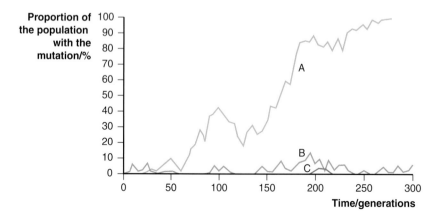

(a)(i) Identify each process and state whether it represents horizontal transmission or vertical transmission. (*4 marks*)

(ii) Describe two ways in which process A differs from mitosis. (*2 marks*)

(b) The overuse of antibiotics may be responsible for the increase in the numbers of bacteria showing resistance to antibiotics. Explain how. (*6 marks*)

(c) Multiple resistance to antibiotics can be acquired by bacteria. Explain how. (*3 marks*)

Total: 15 marks

4 Mutations can be advantageous, harmful or neutral. The graph shows the changes in frequency of individual organisms possessing some mutations over a number of generations. The environment remained constant for the duration of the observations.

(a) What is a mutation? (*2 marks*)

(b) Which of the three lines A, B, or C represents a harmful mutation, which represents a neutral mutation and which represents an advantageous mutation? Use information from the graph to justify your answers. (*6 marks*)

(c) If the environment had changed after 150 generations, the lines for mutations A and C might have been different. Suggest how and explain your answer. (*3 marks*)

Total: 11 marks

Chapter 15

Putting organisms in their places: classification

This chapter covers:
- artificial and natural classifications
- the species concept
- naming organisms
- the five kingdom classification
- phylogenetic trees
- DNA hybridisation
- protein analysis by immunology
- other systems of classification

We all put things into groups. For example, we classify other people into likeable and not so likeable, and gardeners recognise some plants as 'weeds'. In maths, we classify shapes into quadrilateral, circles and so on. We might then subdivide quadrilaterals into squares, rectangles and other shapes.

Classifying things helps us to make sense of the world. Because we believe a person to belong to a certain group (say, our friends) then we know that the person will have a particular set of features. If someone describes a person to you as 'my friend' then, without much more information, you can understand why those two people behave towards each other the way they do. If someone describes a plant as a weed, then you understand why they want to get rid of it from the garden – because you have an understanding of what to be a 'weed' means.

However, these classifications are based on fairly superficial features. One person's friend may be someone else's enemy. A weed to one gardener may be a crop plant to another. They are **artificial classifications**. An artificial classification of animals could put a tarantula and a cobra in the same group because they are poisonous, even though the tarantula is a spider and the cobra is a reptile.

Helmut Partsch/SPL

Weed or crop plant?

Dandelions grow quickly and compete with many other plants for resources. Their leaves can be used in salads.

Saveria Upcraft

The same group or different?

Both animals are venomous, but one is an arachnid and the other is a reptile.

Patrick Fox

Biologists are interested in **natural classifications**. These should not be based on superficial features and ought not to be the subject of disagreement. Natural classifications attempt to reflect the way in which different groups are thought to have evolved. Therefore, groups that went their separate evolutionary ways billions of years ago are unlikely to be closely related and will be placed in groups that reflect this; groups that diverged from each other only a few million years ago (quite recently in evolutionary time!) are much more closely related and, again, their classification reflects this.

Putting organisms into groups that reflect their evolutionary history ◀ is called phylogeny.

Box 15.1 Why classify organisms?

Natural classifications help you to make sense of the world. I could say, 'many of the organisms that have eukaryotic cells with cellulose cell walls, vacuoles and chloroplasts and have a body made up of a root, stem, leaves and flowers are used as crops'. Or I could say, 'many plants are used as crops'. Because you know that the group 'plants' has certain attributes, you know exactly what I am talking about.

What do natural classifications classify?

We cannot possibly observe and classify every single organism. So, we must make some assumptions. We assume, for example, that although there is some variation between them, one African cheetah is very much like another.

A normal cheetah (a) and a 'king' cheetah (b)

The difference in the pattern of spots is not enough for us to believe that either animal is anything but a cheetah.

Perhaps we assume the same about humans, oak trees, salmon and trout. However, with the exception of humans, we shouldn't. There are different types of oak tree, salmon and trout. Aren't there different types of humans? We aren't all alike – for example, people from Wales are different from people from Korea. So how different is different?

The word 'type' does not help — it could mean anything. If we replace it with the biological term **species**, then we are in a better position to explain. There are different species of oak, salmon and trout, but all humans belong to the same species. So our assumption is that all members of the same *species* are fundamentally the same. All 'cork oak' trees are very similar, but different enough from 'English oak' trees to be a different species. This, however, takes us back to the same question: 'how different is different?'

What we need is a working definition of a species. The Institute of Biology defines a species as:

> A species is an interbreeding group of organisms that produces viable and fertile offspring which share a common ancestry and are similar in anatomy and biochemistry.

So, there is the answer. A man from Korea can marry a woman from Wales, they can have children and live happily ever after. However, there will be no acorns on an English oak if it is pollinated by a cork oak.

Now, we can think about classifying organisms. We assume that all organisms in the same species are effectively 'the same' and we place each species into larger groups of several species, and then place those larger groups into even larger groups and so on. It's rather like Russian dolls, except that each Russian doll contains just one smaller doll whereas a classification group contains several groups of the 'next size down'.

◀ Viable means capable of healthy life and fertile means capable of producing offspring. If an organism is fertile, it probably goes without saying that it is viable, so fertility of the offspring is the key.

What do we call the different classification groups?

The groups are collectively called **taxa** and the study of classification is called **taxonomy**.

◀ Taxa is the plural of taxon.

There are different levels of taxon. The lowest level is the species, which groups together all organisms of a single type that are capable of interbreeding to produce fertile offspring.

Species that are very similar to each other are placed in a larger group called a **genus**. For example, all the small cats are placed in the genus *Felis*. Within that group, there are several 'cats', including the domestic cat, *Felis catus*, the European wild cat, *Felis silvestris*, and the lynx, *Felis lynx*.

In fact, some species are divided into subspecies. These are groups within a species that show ◀ some differences and often inhabit different geographical areas. There are two subspecies of the African cheetah; one is found mainly in southern Africa and the other mainly in eastern Africa. However, for our purposes, we can regard the species as the smallest group.

Domestic cat, wild cat and lynx

Notice the double-barrelled scientific name. It is called a **binomial**. All small cats have the same 'first name'. This is the genus to which they belong (*Felis*). It is their **generic** name. The last name is different for each cat and it is this that identifies the kind of *Felis* (cat) it is. All organisms are named in this way. Humans are *Homo sapiens*. Homo is the genus 'Man' and *sapiens* tells us that we are 'intelligent man' — although our treatment of the planet does make you wonder about the validity of this name.

◀ When used to describe the genus, the term 'man' does not imply gender.

Besides the cats mentioned above, there are other 'cat-like' mammals such as lions, tigers, cheetahs and leopards. The name of the lion is *Panthera leo*. It is in the genus *Panthera* (the panthers). The cheetah, *Acinonyx jubatus*, is from a different genus. However, because these big cats and the smaller cats are quite similar, they are all placed into one larger group, the **family** of cats — the Felidae.

Cats are clearly different from dogs, yet there *are* similarities. They are both carnivorous mammals and so are grouped into the **order** Carnivora within the **class** Mammalia. Mammals belong to the **phylum** Chordata, which includes all the vertebrates. Finally, chordates belong to the **kingdom** Animalia — the animals.

Box 15.2 Arthropods

Another animal phylum is the arthropoda. These are invertebrate animals that have an exoskeleton and jointed limbs.

The main classes of arthropods are the insects, crustaceans, arachnids and myriapods.

◀ The name arthropoda gives away the key feature of this group. Think of the first part of the word, *arthro-*; it implies joints, as in *arthr*itis. Poda are limbs, so arthropoda have jointed limbs.

(a)

Damselfly — an insect

(b)

Woodlice — crustaceans

(c)

Scorpion — an arachnid

(d)

Centipede — a myriapod

The kingdom Animalia is one of five kingdoms used in one current system of classification. In the not too distant past, there were other systems, one with four kingdoms and one with two kingdoms. Most of the evidence for each of these systems is based on anatomy and this has led to some anomalies. The protoctistans in particular may be subdivided in the future because this kingdom contains a diverse mix of organisms, from giant kelp to the humble amoeba. Some biologists believe that there may be 20 kingdoms accommodated in this one group. At the moment, it is a kind of taxonomic dumping ground. If an organism does not obviously belong to one of the other kingdoms, then 'it must be a protoctistan!'

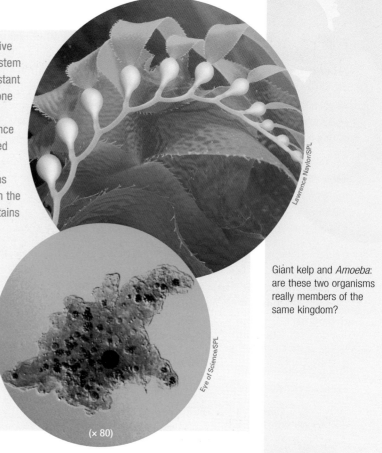

Lawrence Naylor/SPL

Eye of Science/SPL

(× 80)

Giant kelp and *Amoeba*: are these two organisms really members of the same kingdom?

What are the five kingdoms?

Kingdom Prokaryotae

Organisms are either unicellular or filamentous (strands of cells).

In all cases:
- cells lack true nuclei
- cells have circular DNA
- cells have no membrane-bound organelles (e.g. mitochondria, chloroplasts)
- cell walls are made of peptidoglycan (not cellulose)

Examples are bacteria, including cyanobacteria (blue-green bacteria).

Figure 15.1 Bacteria cells

◀ Organisms in the other four kingdoms have eukaryotic cells with:
- true nuclei
- chromosomes
- membrane-bound organelles (e.g. mitochondria and chloroplasts)

Bacterial cell

DNA

Cell wall

10 μm

Filament of blue-green bacteria

10 μm

Kingdom Protoctista

This group is very diverse, and membership is often by exclusion from all other groups.

All protoctistans have eukaryotic cells. Some possess cell walls (non-cellulose), chlorophyll and can photosynthesise; others have no cell walls and are motile. Some are unicellular; others are composed of billions of cells. Examples include *Amoeba* and *Laminaria* (a giant brown seaweed) (see the photographs on page 266).

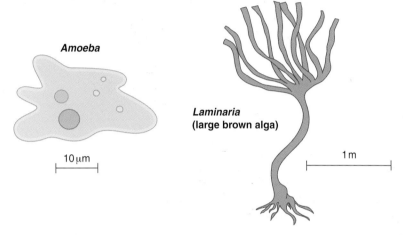

Figure 15.2 *Amoeba* and *Laminaria*

Kingdom Fungi

Fungi are also a diverse group, but have some common features, including:
- a non-cellulose cell wall (it is often made from chitin)
- they are non-photosynthetic
- eukaryotic cells (although filaments, or hyphae, are often multinucleate and not divided into separate cells)
- they secrete enzymes to digest organic materials (usually dead, but some feed from living hosts) outside their cells and absorb the products of digestion

Examples include *Mucor*, yeast and the shaggy ink-cap.

Figure 15.3 Fungi

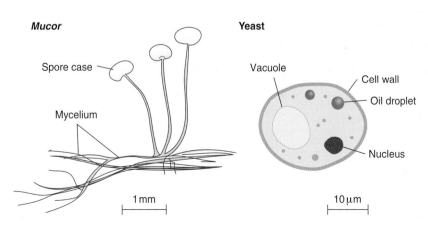

Kingdom Animalia

All animals:

- are multicellular
- have eukaryotic cells with no cell walls
- develop from a blastocyst (a hollow ball of cells)

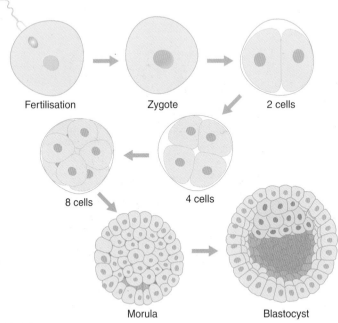

Fertilisation → Zygote → 2 cells → 4 cells → 8 cells → Morula → Blastocyst

Figure 15.4 Formation of a blastocyst

Most animals:

- ingest their food into a digestive system
- are motile

Examples include human beings, domestic cats and basking sharks.

A basking shark

Dan Burton/Alamy

Kingdom Plantae

All plants:

- are multicellular
- have eukaryotic cells with a cellulose cell wall
- are photosynthetic

Examples include mosses, ferns, conifers (plants with cones and needle-like leaves) and angiosperms (flowering plants).

In forests such as this, many different types of plant can be found in a small area

Viruses are not included anywhere in this classification system as they are **acellular** — they are not made of cells. They are often referred to as microorganisms because of their submicroscopic size. All viruses are pathogenic (cause disease) because they have no organelles. They invade living cells and direct the organelles of the infected cell to manufacture new viruses.

Where does the common ancestry and biochemistry fit into classification systems?

All the features used so far to identify the kingdoms and other groups are based on anatomy. For many hundreds of years, these were the only features available. However, the advent of biochemistry, coupled with an understanding of genetics, made available other ways of inferring relationships and other tools to test those inferences. Surely the DNA from different members of the same species should be very similar? Since DNA codes for proteins, won't the proteins also be similar? If the DNA (or proteins) of two species is only slightly different, aren't they more closely related than two species whose DNA (or proteins) shows more differences? Using these two ideas, we can construct a 'phylogenetic tree' which shows:

- how closely related species are
- the 'best guess' using mutation rates and other evidence as to when different species 'diverged' from each other

Figure 15.5 shows such a tree for the baleen whales. It was constructed using evidence from mitochondrial DNA and the sequence of amino acids in one of their cytochrome proteins.

Cytochrome proteins are important in aerobic respiration

Figure 15.5 Phylogenetic tree of baleen whales

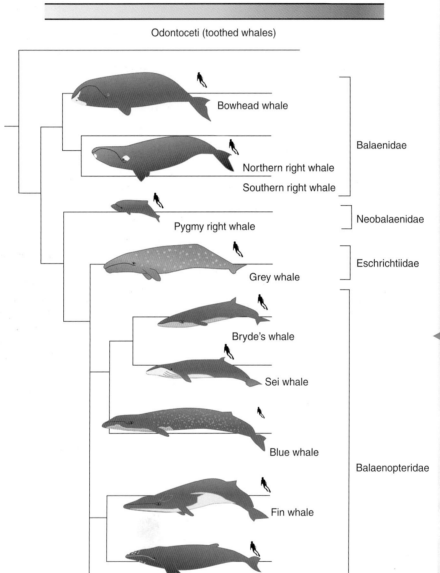

30   25   0   15   10 million years ago

Odontoceti (toothed whales)

Bowhead whale

Northern right whale

Southern right whale

Balaenidae

Pygmy right whale

Neobalaenidae

Grey whale

Eschrichtiidae

Bryde's whale

Sei whale

Blue whale

Balaenopteridae

Fin whale

Humpback whale

Antarctic minke whale

North atlantic minke whale

Baleen whales filter water through sheets of baleen to obtain their food. The toothed whales are called odontoceti.

How do we find out how similar one DNA is to another?

Ideally, we would compare all the genes in each species for their precise base sequence. However, this would take far too long. Instead, we use a technique called **DNA hybridisation**. The technique measures the extent to which a strand of DNA from one species can bind with (hybridise with) a strand of DNA from another species. The main steps in the process are:

- Extract the total DNA from the nucleus of a cell of each species (say, species A and species B).
- Heat each sample of DNA until it separates completely into single strands (ssDNA).
- Remove all the non-coding DNA.
- Radioactively label some of the ssDNA from species A.
- Cool. Allow some of this radioactive ssDNA to hybridise with non-radioactive ssDNA from species A in one tube and with non-radioactive ssDNA from species B in a second tube.
- Reheat gradually, taking regular samples to estimate the percentage of DNA strands that have separated again. This is found from the amount of radioactivity in a sample.
- Draw a graph showing the percentage of ssDNA in each sample and determine the temperature at which 50% of the DNA has separated into single strands $T_{50}H$.
- The information can then be used to calculate the percentage similarity of the two DNA samples.

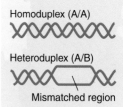

Homoduplex (A/A)

Heteroduplex (A/B)

Mismatched region

Figure 15.6
Hybridisation of DNA

◀ The aim is to compare the DNA found in the genes, so the non-coding DNA (Chapter 10, page 175) must be removed.

◀ Although the ssDNA from species A and species B will hybridise (bind) it will not do so along all its length. There will be regions that are 'mismatched' (the base pairs are not complementary) and so do not bind. It is, therefore, easier to separate these strands than strands that are fully complementary.

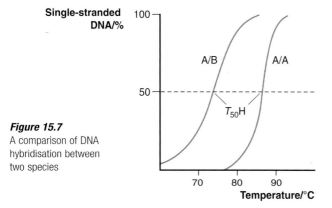

Figure 15.7
A comparison of DNA hybridisation between two species

From DNA hybridisation techniques, the phylogenetic tree of humans and the great apes appears as in Figure 15.8.

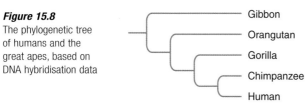

Figure 15.8
The phylogenetic tree of humans and the great apes, based on DNA hybridisation data

Gibbon
Orangutan
Gorilla
Chimpanzee
Human

How do we find how similar one protein is to another?

Here, we are talking about the similarity between a single protein found in several species, not *all* the proteins. So, we might compare, for example, how similar human haemoglobin is to haemoglobin from a chimpanzee or a goldfish. We would expect to find that our haemoglobin is more similar to that of a chimp than a goldfish. How do we do this? We could, ideally, determine the amino-acid sequence of each sample of protein, and in some cases (e.g haemoglobins) this is done. Generally, however, that takes far too long. Instead we rely on the immune system's reaction to 'foreign' (non-self) protein (Chapter 8). If a protein from another organism enters your body, it will be 'attacked' by the immune system. Antibodies will be produced against it and the protein will be destroyed or precipitated.

In this technique, a protein from species A is injected into an experimental animal. The animal produces antibodies against the protein. Shortly afterwards, the same protein from species B is exposed to the antibodies and the strength of this reaction is estimated. If the proteins are very similar, the second reaction will be stronger than if they are only slightly similar.

From data obtained by this immunological technique, the phylogenetic tree of humans and the great apes appears slightly different from that obtained using data from DNA hybridisation.

Figure 15.9 Phylogenetic tree of humans and great apes based on immunological data

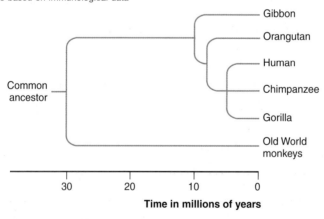

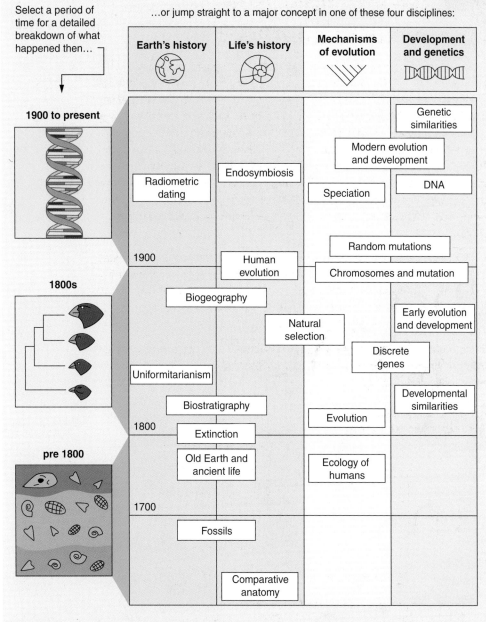

Select a period of time for a detailed breakdown of what happened then…

…or jump straight to a major concept in one of these four disciplines:

| Earth's history | Life's history | Mechanisms of evolution | Development and genetics |

1900 to present

1900

1800s

pre 1800

- Genetic similarities
- Modern evolution and development
- Endosymbiosis
- Radiometric dating
- Speciation
- DNA
- Random mutations
- Human evolution
- Chromosomes and mutation
- Biogeography
- Natural selection
- Early evolution and development
- Uniformitarianism
- Discrete genes
- Biostratigraphy
- Developmental similarities
- Evolution
- Extinction
- Old Earth and ancient life
- Ecology of humans
- 1700
- Fossils
- Comparative anatomy

Figure 15.10 Changing ideas and new evidence over the past 300 years have influenced our thinking on evolution and classification

Biochemical evidence sometimes leads us in a different direction from the evidence based on anatomy and development. However, it can create a real headache. Based on this kind of evidence, the phylogenetic tree for salmon, lungfish and cows is shown in Figure 15.11.

According to this, the lungfish (usually placed in the class 'Pisces') is more closely related to the cow (class Mammalia) than it is to the salmon (class Pisces)!

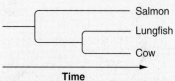

Salmon
Lungfish
Cow

Time

Figure 15.11 Phylogenetic tree of salmon, lungfish and cow based on biochemical evidence

Recent research has shown that the genetics and metabolism of anaerobic bacteria are completely different from the genetics and metabolism of other organisms. This has led to another classification system based on the idea of three 'domains', with an uncertain number of kingdoms (Figure 15.12). Table 15.1 compares the old two-kingdom system, the five-kingdom system and the three-domain system.

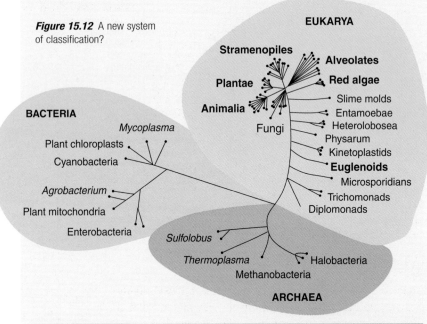

Figure 15.12 A new system of classification?

Table 15.1 The classification systems compared

Two kingdom system	Five kingdom system	Three domain (and who knows how many kingdoms) system	
Animalia	Animalia		Animalia
Plantae	Fungi		Fungi
	Plantae		Plantae
either Protozoa (= Animal) or Algae (= Plant)	Protoctista	Eukarya	Alveolata
			Stramenopiles
			etc…
			Sporozoa
			Mycetozoa
			etc…
			Archezoa
Bacteria and blue-green algae (= Plant)	Prokaryotae	Eubacteria	(kingdoms not specified)
		Archaea	Euryarchaeota
			Korarchaeota
			Crenarchaeota

Why do mitochondria and chloroplasts appear in a classification system?

According to the endosymbiont theory, the mitochondria and chloroplasts of modern plant cells were once free-living prokaryotic cells. The mitochondria were a type of aerobic bacterium; the chloroplasts were a type of cyanobacterium (blue-green bacterium). This is summarised in Figure 15.13. There is some genetic and other evidence to support this.

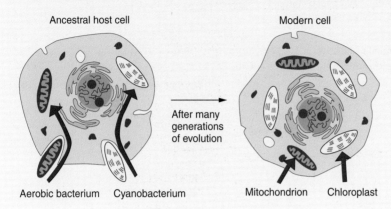

Figure 15.13 The endosymbiont theory of the origin of mitochondria and chloroplasts

How natural is our natural classification system?

There are many questions to be answered here. The species is probably a fairly natural group. After all, there are many groups of organisms with anatomical and biochemical similarities that do interbreed to produce fertile offspring.

What about organisms that reproduce asexually? If they do not breed with another organism, how can we know whether or not they are the same species? What about the occasions when fertile hybrids occur?

> A hybrid is the offspring of two different types, sometimes of two different species.

Production of hydrids happens more frequently than you might think. The wheat that is used today to make bread and pasta is the fertile hybrid of a fertile hybrid!

Even so, the species is the most natural of the taxa. After all, who decided that one group of big cats should be in the genus Panthera and another group should be in the genus Acinonyx? Most of the other taxa are, by comparison to the species, quite arbitrary. So, is it just another artificial system after all…?

Summary

- Artificial classifications use superficial and subjective criteria; natural classifications use features that are objective and that attempt to show evolutionary relationships between different groups.
- The main taxa (from largest to smallest) are: kingdom, phylum, class, order, family, genus and species.
- A species is a group of organisms that breed to produce viable and fertile offspring, share a common ancestry and are similar in anatomy and biochemistry.

- The five kingdoms are Prokaryotae, Protoctista, Fungi, Plantae and Animalia.
- DNA hybridisation can produce estimates of the similarity of DNA from two different species; this can be used to infer how closely related the two species are.
- Protein sequencing and protein immunology can be used to compare how similar particular proteins from different species are.
- The results from DNA hybridisation and protein immunology can be used to construct phylogenetic trees that show the relationships and likely evolutionary origins of different species; these techniques sometimes produce results that are in conflict with currently established classifications.
- There are other classification systems that propose different organisation of kingdoms.
- Although the species is a relatively natural group, some of the other divisions are more arbitrary.

Questions

Multiple-choice

1 A species is an interbreeding group that produces offspring that are:
   A viable but not fertile
   B fertile but not viable
   C both viable and fertile
   D neither viable nor fertile

2 The order of taxa from largest to smallest is:
   A species, genus, family, order, class, phylum, kingdom
   B species, genus, family, order, class, kingdom, phylum
   C kingdom, phylum, class, order, family, species, genus
   D kingdom, phylum, class, order, family, genus, species

3 The kingdom that contains organisms, all of which have eukaryotic cells, some of which have a non-cellulose cell wall, some of which can photosynthesise and some of which are motile is:
   A animalia
   B plantae
   C protoctista
   D fungi

4 The difference between artificial and natural classifications is:
   A artificial classifications rely on only superficial features
   B natural classifications rely on features that show evolutionary relationships
   C neither A nor B
   D both A and B

5 Plants are different from animals because:
   A plants have prokaryotic cells whereas animals have eukaryotic cells
   B plant cells have cellulose cell walls whereas animal cells have non-cellulose cell walls
   C only animals form a zygote at fertilisation
   D plants do not develop from a blastocyst

6 The cheetah *Acinonyx jubatus* belongs to:

 A the genus *Acinonyx*

 B the species *Acinonyx jubatus*

 C neither A nor B

 D both A and B

7 In DNA hybridisation, the similarity between DNA of two species is determined by:

 A the extent to which strands of the different DNA bind on heating

 B the extent to which strands of the different DNA bind on cooling

 C the extent to which strands of the different DNA separate on heating

 D the extent to which strands of the different DNA separate on cooling

8 A phylogenetic tree of four organisms is shown below

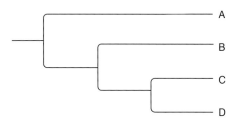

The phylogenetic tree shows that:

 A species A is most closely related to species D

 B species C is most closely related to species B

 C species C and D are related equally to species A

 D species A is related most distantly to species D

9 Fungi have:

 A eukaryotic cells with cellulose cell walls and no chloroplasts

 B prokaryotic cells with cellulose cell walls and no chloroplasts

 C prokaryotic cells with non-cellulose cell walls and no chloroplasts

 D eukaryotic cells with non-cellulose cell walls and no chloroplasts

10 The theory behind using protein biochemistry to classify organisms is that:

 A the more similar the proteins, the more closely related the organisms

 B the degree of similarity between proteins measures how long ago species diverged

 C similar organisms have similar proteins because they have similar DNA

 D all of the above

Examination-style

1 The diagram shows the distribution of two species of animals living on an island.

(a) Explain what is meant by the term 'species'.

 (2 marks)

(b) The territories of the two species overlap in region X, yet there are no fertile hybrids found in this region.

 (i) What is a fertile hybrid? *(2 marks)*

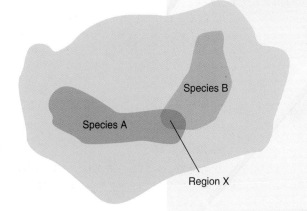

Species B

Species A

Region X

(ii) How does the absence of fertile hybrids support the hypothesis that these are in fact two different species? (*2 marks*)

Total: 6 marks

2 *Amoeba proteus* and *Paramecium caudatum* are both unicells. They both have eukaryotic cells and are motile.

(a)(i) Copy and complete the table showing the classification of *Amoeba*.

Taxon	Taxonomic group
Kingdom	
	Plasmodroma
	Rhizopoda
	Amoebida
	Amoebidae
Genus	
Species	

(*2 marks*)

(ii) Name **three** taxa that the two organisms could share. (*2 marks*)

(iii) Name **one** taxon that the two organisms do not share. (*1 mark*)

(b) Give **two** differences between these cells and bacterial cells. (*2 marks*)

Total: 7 marks

3 The amino acid sequences of one of the polypeptide chains of haemoglobin from nine animals were determined. The results are shown in the table.

Type of haemoglobin	Number of amino acids different from human haemoglobin
Human	0
Gorilla	1
Gibbon	2
Rhesus monkey	8
Horse	25
Kangaroo	38
Chicken	45
Frog	67
Sea slug	127

(a) Using the information in the table, draw a phylogenetic tree of the organisms. (*4 marks*)

(b) It is possible to use protein biochemistry to suggest relationships between species. Explain why. (*4 marks*)

(c) Immunological determination of protein similarity is often preferred to amino acid sequencing. Explain why. (*2 marks*)

Total: 10 marks

4 'The species is the basis of our classification system. It is the most natural of all the taxa.'

(a) Explain what is meant by the term 'species'. *(2 marks)*

(b) Give *one* example of a situation in which the usual definition of a species does not quite work. Explain why. *(3 marks)*

(c) Choose *one* of the other taxa and suggest why we may consider it to be less natural than the species. *(2 marks)*

Total: 7 marks

Chapter 16

Variation in the environment: biodiversity

This chapter covers:
- the nature of biodiversity: species richness or species diversity
- how to calculate and interpret an index of diversity
- the impact of deforestation on biodiversity
- the impact of agriculture on biodiversity

Rainforest is being cut down at an alarming rate, and we are losing species of plants and animals that we don't even know exist. These organisms form part of the natural wealth of living things, not just of the rainforest but of the whole planet. Replacing rainforest with agricultural crops or a single type of tree cannot replace the huge mass of interrelated plants, animals and other organisms that inhabited the area. The biodiversity of the planet is being reduced minute by minute.

What is biodiversity?

So far, we have covered the variation within and between different species and other groups of organisms. Now, we look at the product of evolution, the variety of organisms found on our planet — its biodiversity. The most usual way to think of biodiversity is in terms of **species richness**. This is the number of different species present in an **ecosystem**.

> An ecosystem is a self-sustaining system in which organisms interact with each other and with their physical environment. An ecosystem can vary in size from a garden pond to a tropical rainforest.

However, if only one or two individuals of a particular species are present in an ecosystem, they will not be contributing a great deal to the biodiversity of the system. A more useful concept is **species diversity**. This takes into account, not just how many species are present, but the success of each species in the ecosystem. An **index of diversity** can be calculated, which can be used to give a picture of the ecosystem as a whole.

Look at the examples below. They are fictitious, but the figures serve to illustrate a point. Each area contains the same six species and the same total number (100) of organisms, yet the areas are clearly very different.

Table 16.1

Number of organisms of each species			
Species	Area 1	Area 2	Area 3
A	86	16	23
B	5	17	25
C	2	16	27
D	3	17	5
E	1	17	12
F	3	17	8

- In area 1, only species A is successful; it dominates the area.
- In area 2, the success of all six species is approximately equal.
- In area 3, species A, B and C dominate the area.

The species diversity of the three areas should reflect the difference in abundance of the six species within each area. **Simpson's index of diversity** is calculated from the formula:

$$d = \frac{N(N-1)}{\Sigma n(n-1)}$$

In this formula, d is the index of diversity, N is the total number of organisms in the area and n is the total number of organisms of each species.

For area 1:
$$d = \frac{100 \times 99}{(86 \times 85) + (5 \times 4) + (2 \times 1) + (3 \times 2) + (1 \times 0) + (3 \times 2)} = 1.348$$

For area 2:
$$d = \frac{100 \times 99}{(16 \times 15) + (17 \times 16) + (16 \times 15) + (17 \times 16) + (17 \times 16) + (17 \times 16)} = 6.314$$

For area 3:
$$d = \frac{100 \times 99}{(23 \times 22) + (25 \times 24) + (27 \times 26) + (5 \times 4) + (12 \times 11) + (8 \times 7)} = 4.911$$

A low value for the diversity index suggests an area dominated by one or just a few species. If there are more successful species with no species dominating the area, the value of the diversity index is higher.

What does the index of diversity tell us about an area?

A low value for the index of diversity suggests the presence of only a few successful species, perhaps only one.

This could be because the environment is hostile and only a few organisms are well-adapted to it. Change in the environment could have quite serious effects. If those species that can survive are seriously affected, then the fledgling ecosystem may be disrupted.

There are other ways of calculating Simpson's index of diversity. This method gives 1 as the lowest possible value, representing an area with just one species. Another method gives 1 as the highest value, representing an area with a large number of successful species.

There might be only a few types of organism in an area because they outcompete other similar types that could survive in that environment. Rhododendron bushes effectively prevent other plants from growing in the same area by shading them so they cannot photosynthesise, and by secreting chemicals into the soil that inhibit the germination of seeds other than rhododendron seeds. This means that if the rhododendrons die, it will be a long time before any other organisms can colonise the area.

Lichens growing on bare rock

There are not many organisms here other than the lichens. The value of the diversity index will be low.

A higher diversity index suggests a number of successful species and a more stable ecosystem. More ecological niches are available and the environment is likely to be less hostile. Environmental change is likely to be less damaging to the ecosystem as a whole, unless it affects all the plants present. A tropical rainforest is an example of a stable ecosystem with high species diversity.

Tropical rainforest

There are many successful species here; the index of diversity is high.

How have humans influenced biodiversity?

Humans have influenced the environment far more than any other species. This is one of the key features of human evolution. We have not so much adapted to the environment by natural selection as changed the environment to suit us. Until relatively recently, this has not been a problem. However, the rate of this change has accelerated with the huge human population increase and the development of technology.

We have reduced biodiversity in many ways, but we will focus on:
- deforestation
- the impact of agriculture

Where have all the trees gone?

Deforestation is carried out for two main reasons:
- to clear land for human activity, such as mining, agriculture or house-building
- to obtain timber to make paper, charcoal, furniture, or as a building material

This used to be tropical rainforest

Tropical rainforest is one of the most complex and species-rich ecosystems in the world. Rainforest covers about 7% of the surface of the Earth and contains 25% of the known species.

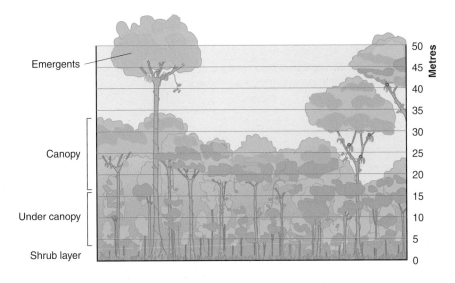

Figure 16.1 Structure of a tropical rainforest.

Although many of the trees are very tall, the root systems are shallow and trees can easily fall. The shallow root systems reflect the shallow, nutrient-poor soils. The soils are nutrient poor because many of the minerals from the soil remain 'locked up' in the huge trees. The only recycling of nutrients that occurs on a regular basis occurs when leaves fall. There is no accumulation of detritus as decomposers rapidly break down the leaves and release the mineral ions they contain. The roots take these up, leaving few mineral ions in the soil. As a result, when the forest is cleared for agriculture, crop yields are often poor after the first year and more forest must be cleared.

Tropical rainforest is the most productive of natural land ecosystems. The net primary production (mass of living tissue produced allowing for losses due to respiration) is 2.2 kg m^{-2} y^{-1}, nearly twice that of temperate forests.

Felling tropical rainforest has far-reaching effects.

- There is a serious reduction in species diversity. Many ecological niches are destroyed when trees are felled and the species that fill these niches are lost.
- There is a reduction in the rate at which carbon dioxide is removed from the atmosphere. In addition, if the trees are burned, then carbon dioxide is added to the atmosphere. The local and global cycling of carbon is therefore affected.
- There is a reduction in the amount of nitrogen returned to the soil. The nitrogen fixed in the proteins and other compounds in the massive tree trunks remains fixed in a teak table in London or New York. Any tree trunks not removed from the area are slow to decay and the soil is depleted in nitrate for many years.
- A secondary succession may take place. If the felled area is allowed to regenerate, then seeds of many species of plant from the shrub layer will germinate. This could give rise to a less complex ecosystem with a lower biodiversity. Ordinarily, the shrub layer is limited because the canopy prevents much light from reaching lower levels.

It is estimated that 20% of the Amazon rainforest has been felled since 1970. This is expected to rise to 40% by 2025, leaving only 60% of the rainforest that ◄ was present in 1970.

The canopy prevents much light from penetrating to the shrub layer

The felling of trees need not be totally destructive and the practice need not be halted. However, the rainforests must be conserved, and felling and re-planting in a planned cycle over a number of years can do this. This could give a sustainable yield of timber, without endangering the species diversity of the rainforests.

Box 16.1 What about that burger?

Before you eat your next quarter-pounder, bear in mind that to make that one burger using South American beef requires:

- the clearing of 5 m² of rainforest
- the destruction of 75 kg of living matter, including 20–30 plant species, up to 100 insect species and dozens of bird, mammal and reptile species

A quarter-pounder

What about the effects of agriculture?

The effects of large-scale agriculture usually follows a similar pattern. Large areas of land are given over to the production of just one crop plant, such as maize or another cereal. This change brings a reduction in biodiversity for several reasons, including:

- The area is dominated by just one species, which drastically reduces the number of niches available for other organisms.
- Organisms that might live there are regarded as pests because they reduce the crop yield; they are controlled by pesticides.
- Hedgerows are removed to create bigger, more productive fields; this further reduces the number of habitats and niches and, therefore, the biodiversity of the area.

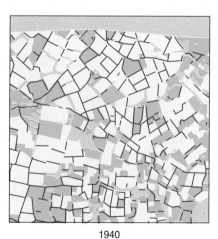

1940

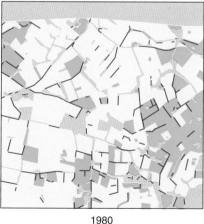

1980

Grassland
Crops
Other
∧ Hedgerow

0   km   1

Figure 16.2 The map shows the change in the amount of agricultural land, grassland and hedgerows between 1940 and 1980 in an area along the Moray Firth in Scotland.

Other changes in agricultural practices have also affected biodiversity. Traditionally, crops were 'rotated'; one year a cereal would be grown, followed the next year by perhaps a root crop such as carrots, the next year by a legume such as beans and maybe one year 'fallow' (just grass, no crop).

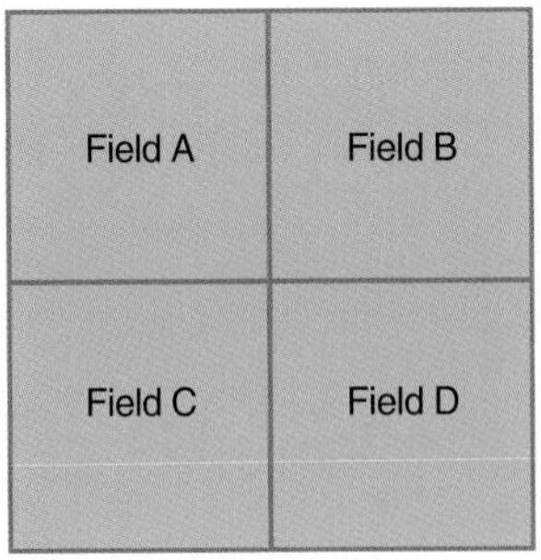

Field A	Field B
Field C	Field D

Field	Crop			
	Year 1	Year 2	Year 3	Year 4
A	Root	Legume	Fallow	Cereal
B	Cereal	Root	Legume	Fallow
C	Fallow	Cereal	Root	Legume
D	Legume	Fallow	Cereal	Root

Figure 16.3 Crop rotation

The rotation was carried out with different timings in different fields, so that all crops were always available. This meant that there were many habitats available for different animals. The intensive farming approach of planting a single crop year after year, and keeping pests at bay with herbicides and pesticides, reduces the habitats available. These practices have been blamed for the decline of the brown hare and the grey partridge in Scotland.

Box 16.2 It's not all doom and gloom

As scientists have found out more about the impact of intensive agriculture, the information has been passed to government at national and local levels. Many local councils are taking matters into their own hands. One example is the action plan drawn up by Falkirk Council for arable land (land cultivated for crops).

The council has recognised the fact that few farmers manage arable land for the benefit of wildlife. They want to make a profit, and understandably so. However, this can be achieved while also increasing the biodiversity of the area. The Falkirk area biodiversity action plan has three main objectives, with a number of smaller targets within each.

Objective 1: to encourage the retention or introduction of spring sowing and over-wintering stubbles

Over-wintering stubbles are the remains of plants that have been harvested. Retaining these allows birds to feed in the winter and some rarer plants to grow. Spring sowing of seeds provides cover for ground-nesting birds.

Objective 2: to increase the area of land that is not harvested or cropped intensively

This allows a wider variety of habitats and so increases biodiversity.

Objective 3: to enhance the extent and quality of cereal field margins; by 2011 there should be at least 30 hectares (75 acres) of these margins

Field margins are areas at the edges of crop fields that provide feeding and nesting sites for a variety of birds. They also support large numbers of invertebrates that are food for the birds. Small mammals live in these margins and they, in turn, attract predators. In addition to increasing the biodiversity of the area, field margins also act as 'buffer zones', preventing the run-off of fertilisers and pesticides into nearby areas.

The Falkirk Council plan is just one local example. It is a good plan because it has set achievable objectives that will have a definite impact. There are other local councils enacting their own action plans, and national governments and international organisations are beginning to take notice of the danger to the biodiversity of our planet. You can do something too — don't forget the quarter-pounder!

Summary

- Species richness is the number of species in an area, irrespective of the number of individuals of each species
- Species diversity takes into account the number of species in an area and the relative success of each species.
- If the numbers of the organisms in an area can be estimated, an index of diversity can be calculated.
- Deforestation reduces biodiversity because:
  - ecological niches are lost when trees are felled
  - regeneration of an area of cleared rainforest may not result in the same trees growing again; a less complex ecosystem with lower biodiversity is more likely
- Intensive agriculture reduces biodiversity because:
  - often, just one crop is sown in the same field year after year; many habitats are lost and the index of diversity is reduced drastically
  - the use of pesticides and herbicides further reduces the biodiversity by killing plants and animals that would inhabit the area
  - the removal of hedgerows further reduces the number of available habitats
- Local, national and international plans can all help to reverse the decrease in biodiversity; personal action can also have an impact

Questions

Multiple-choice

1 Species richness takes into account:

  A the number of different species

  B the numbers of each species

  C both A and B

  D neither A nor B

2 A low index of diversity could indicate:

  A an unstable ecosystem with many niches

  B a stable ecosystem with many niches

  C a stable ecosystem with few niches

  D an unstable ecosystem with few niches

3 Deforestation could lead to:

  A lower biodiversity because of loss of habitats

  B reduced uptake of carbon dioxide from the atmosphere

  C a less complex ecosystem on regeneration

  D all of the above

4 The maintenance of field margins in cereal fields improves biodiversity because it:

  A creates a habitat for many invertebrates

  B creates nesting sites for birds

  C provides a habitat for small mammals

  D all of the above

5 Which of the following agricultural practices would be most likely to increase biodiversity?

A crop rotation

B cultivation of just one crop

C the use of pesticides

D not sowing crops in spring

Examination-style

1 The table gives information about the area of Brazilian Amazon rainforest lost per year and the area of forest remaining.

Year	Area of forest /km²	Annual forest loss/km²	Percentage of 1970 cover left	Forest lost since 1970/km²
1970	4 100 000		100.0	
1977	3 955 870	21 130	96.5	144 130
1987	3 744 570	21 130	91.3	355 430
1990	3 692 020	13 730	90.0	407 980
1997	3 576 965	13 227	87.2	523 035
2001	3 505 932	18 165	85.5	594 068
2004	3 432 147	27 429	83.7	667 853
2006	3 400 254	13 100	82.9	699 746

(a) Plot a graph of percentage cover left against time. Extend your graph to the year 2020. *(5 marks)*

(b) Use your value for 2020 to calculate the area of Amazon rainforest that will be left by then. *(1 mark)*

(c) Suggest why, in 2004, people were particularly alarmed by the figures for annual loss. *(3 marks)*

(d) Explain how the loss of rainforest reduces biodiversity. *(6 marks)*

Total: 15 marks

2 In an investigation into two areas, the numbers of several types of plant were recorded. The results are shown in the table below.

Plant species	Number of plants (*n*) in area A	Number of plants (*n*) in area B
Woodrush	4	2
Holly	8	3
Bramble	3	3
Yorkshire fog	6	12
Sedge	7	4
Buttercup	7	4
Total (*N*)	35	28

(a) Use the formula

$$d = \frac{N(N-1)}{\Sigma n(n-1)}$$

to calculate the diversity index for each area. *(3 marks)*

(b) Give three deductions that might be made about an area with a low diversity index. *(3 marks)*

(c) Explain *two* ways in which agricultural practices can reduce biodiversity. *(4 marks)*

Total: 10 marks

3 Read the following statements about tropical rainforest.

- Experts estimate that deforestation results in the loss of 137 different species (some plant, some animal, some microorganisms) every day.
- More than half the world's 10 million species of plants and animals live in rainforest; this represents a vast genetic store.
- More than 20% of the world's oxygen is produced in the Amazon rainforest.
- A 10-hectare area of rainforest may have a greater species diversity of trees than all of North America.
- The same area contains 9000 tonnes of living plant material.
- Many plants in the rainforest produce chemicals that might be used to treat serious conditions.
- The profit from 1 hectare of rainforest land is:
  - $150, if used for cattle ranching
  - $1000, if the timber is harvested
  - $6000, if renewable resources (nuts, oil-producers, medicinal plants) are harvested

The formula for Simpson's index of diversity is:

$$d = \frac{N(N-1)}{\Sigma n(n-1)}$$

where N = the total number of organisms in an area

n = the total number of organisms of one species in an area

Σ = the sum of

(a) Use the formula and the information given to explain why deforestation will reduce the diversity index of the rainforest. *(3 marks)*

(b) Deforestation leaves a soil that is nutrient poor. Explain how this happens. *(5 marks)*

(c) Use the information given to make a case to a politician to push for a reduction in deforestation. *(7 marks)*

Total: 15 marks

Index